CW00622087

Morphometrics for the Life Sciences

Recent Advances in Human Biology

Series Editor: **Charles E. Oxnard** (*The University of Western Australia*)

Recent Advances in Human Biology – Volume

Series editor: Charles E. Oxnard
Centre for Human Biology
The University of Western Australia

Morphometrics for the Life Sciences

Pete E. Lestrel
University of California, Los Angeles

W⊖ World Scientific
Singapore • New Jersey • London • Hong Kong

Published by

World Scientific Publishing Co. Pte. Ltd.

P O Box 128, Farrer Road, Singapore 912805

USA office: Suite 1B, 1060 Main Street, River Edge, NJ 07661

UK office: 57 Shelton Street, Covent Garden, London WC2H 9HE

British Library Cataloguing-in-Publication Data
A catalogue record for this book is available from the British Library.

ISBN 981-02-3610-7

Printed in Singapore.

PROLOGUE

One aim of this volume is to introduce the subject matter of a comparatively new discipline, but one with ancient roots: morphometrics. Morphometrics is the quantitative description or measurement of the biological form. It represents the initial step in the elucidation of taxonomic and classification questions, in dealing with issues of structure and function, as well as providing the raw material for explanations of biological process. Thus, morphometrics is central to all the biological sciences. Morphometrics deals with the need to quantify, in a complete and precise manner, the visual information inherent in all biological organisms. While this visual information is readily apparent, it has been inexorably difficult to adequately characterize in numerical terms.

As a starting point for the study of morphometrics, when measurement is considered, it is the universally used conventional metrical approach (CMA), consisting of distances, angles and ratios that are immediately thought of. That CMA represents a very inefficient procedure for dealing with the complex and irregular forms that constitute a majority of biological organisms on earth is generally not recognized. Accordingly, a re-evaluation of the way such metrics continue to be used is imperative and has spawned a number of alternative methods over the last three decades. While a complete morphometric description of biological organisms is not yet possible, the appearance of these new approaches can be considered a promising endeavor for continuing advances in morphometrics as the 21st century progresses.

Another aim of this work is to continue in a more generalized setting, the application of Fourier Descriptors (FDs), Fourier Transforms (FTs) and wavelets to characterize the boundary outline of biological forms, which is taken up in Chapter 9. Part of this material represents: [1] a continuation of earlier research efforts and [2] an expansion of ideas initially formulated in an edited volume entitled: *Fourier Descriptors and their Applications in Biology* (1997).

Part one, *Theoretical Background*, consisting of six chapters is intended as an introduction to morphometrics. Chapters 1 and 2 introduce morphometrics and research methods respectively. These are followed by Chapter 3, which contains a brief history of measurement, which leads into Chapter 4, which describes the increasing use of quantification in biology. The last two chapters are more theoretical in nature and deal with a host of issues that impinge directly and indirectly on the quantitative analysis of the biological form. These include topics such as complexity, systems and models (Chapter 5) and the development of a formal model of form (Chapter 6).

Part two, *Morphometric Techniques*, is composed of four chapters followed by an epilogue. These chapters are more applied in character and deal with the numerous and quite diverse morphometric procedures that have been applied to biological datasets. Chapter 7 deals with multivariate morphometrics, Chapter 8 with coordinate morphometrics, Chapter 9 with boundary morphometrics and Chapter 10 with structural morphometrics. A number of sources are provided that allow access to available software that can be used with the methods described in Chapter 7 through Chapter 10. Additionally, more sources are available via the Internet and the student will find that a diligent search will be rewarded.

Most of the material in this volume lends itself for a one-semester interdisciplinary course on morphometrics. It is presented at an introductory level, although extensive references are provided for the inquisitive student who wants to probe more deeply. While it is aimed at the biologically oriented student, it may also be profitably used by students in other related disciplines. An introductory course on morphometrics has been given to post-graduate students at the School of Dentistry, University of California at Los Angeles.

Various parts of the text have been presented at lectures given to the faculty and students at the University of Glasgow Dental School; University of Adelaide, Australia; The University of Otago, Dunedin, New Zealand; The Karolinska Institute, Stockholm; Tokyo Metropolitan University, Hachioji, Tokyo; Tokyo National Science Museum, Ueno, Tokyo; Nihon University School of Dentistry at Matsudo, Chiba; The National University of Singapore; and the University of Bourgogne, Dijon.

Pete E. Lestrel
Van Nuys, California
February 2000

CONTENTS

LIST OF FIGURES

LIST OF TABLES

LIST OF TABLES

ACKNOWLEDGMENTS

The idea for this volume was in the back of my mind even before I had completed *Fourier Descriptors and their Applications in Biology* for Cambridge Press. However, its birth, so to speak, was a chance encounter, a consequence of my visiting Professor Charles Oxnard in Perth, Australia in 1996. Charles suggested I should consider writing it as a volume for the *Recent Advances in Human Biology* series of which he was Editor. Thus, *Morphometrics for the Life Sciences* was born. I am, needless to say, exceedingly grateful to Charles for that suggestion.

I am most indebted to Professor Fumio Ohtsuki, Tokyo Metropolitan University (TMU) for always being a gracious host starting with my first visit to Tokyo in 1995. In addition, for allowing me time in August of 1996 to initiate the first rough draft of the first two chapters of this volume, as well as having me return as Visiting Professor for three months in the summer of 1999 to supervise students and continue our joint research projects on human cranium shape changes and human growth in stature. I also wish to acknowledge the friendship of Professors Hideyuki Tanaka and Teruo Uetake, in whose company I was able to finally climb Mt. Fuji. I also want to mention my relationship with Dr. Osamu Takahashi, Nihon University School of Dentistry at Matsudo, who visited me for 15 months (1997-1999) to carry out a series of joint research projects dealing with human craniofacial complex. Finally, I want to indicate that I feel particularly honored by the gracious treatment I received from Dr. Shigeo Otake, Dean of the School, Professor Eisaku Kanasawa and Dr. Kazutaka Kasai, all from the Nihon University School of Dentistry where I delivered an invited lecture in June 1999.

I would like to thank Professor W. J. S. Kerr, University of Glasgow, Department of Orthodontics, for inviting me to beautiful Scotland and to participate in joint research on functional appliance therapy. Thanks must also go to Dr. Jan Huggare, Chairman of Orthodontics, University of Karolinska, for inviting me to come to Stockholm to give a lecture as well as allowing me the privilege to attend the Nobel Lecture in Physiology/Medicine for 1996 given to Professors Peter C. Doherty and Rolf M. Zinkernagel for their work concerning the specificity of cell-mediated immune defenses. Moreover, Jan was also instrumental in arraigning my 1999 lecture at the National University of Singapore when he was there for a year's appointment.

I am most grateful to Professor Eric Verrecchia, University of Bourgogne, Dijon, for inviting me to the Bio Geo Images '99 conference held in September 1999. Eric and I not only shared a fondness for the food and wine of the Burgundy region, but also a mutual admiration for the exceptional accomplishments of J. B. J. Fourier. I also wish to thank Dr. Hervé Drolon, University of Le Havre, for sending me reprints and for her helpful suggestions with respect to wavelets.

A debt of gratitude goes to Professor Neal Garrett, UCLA School of Dentistry for reviewing parts of the manuscript. I have heavily relied on his advice in the past so I am again grateful for his efforts. I am also especially grateful to Charles Wolfe for his unflagging long-term programming efforts, valuable and stimulating discussions, and for the thorough review of the manuscript. Thanks must go to Dr. Albert Bodt for taking the time out of a busy medical practice at Kaiser Permanente, to review the manuscript; and for his enthusiasm and close friendship for almost twenty years now. Any errors remaining, however, are solely my own.

Lastly and most certainly not least, I owe my wife Dagmar a great deal, not only whose talents as a librarian were particularly helpful, but also for being my soul mate in marriage for the last thirty-two years and who had to tolerate my many late nights and early mornings necessary to bring this project to completion.

PART ONE:
THEORETICAL BACKGROUND

1. INTRODUCTION TO MORPHOMETRICS

The study of form may be descriptive merely, or it may become analytical. We begin by describing the shape of an object in simple words of common speech: we end by defining it in the precise language of mathematics; and the one method tends to follow the other in strict scientific order and historical continuity.

Morphology and mathematics (1915)
D'Arcy Thompson (1860-1948)

1.1. INTRODUCTION

The idea of form is one of the most fundamental concepts underlying all of the sciences. All forms consist of a large number of shared aspects that include size, shape, color, patterning, etc. In the most basic sense, the human ability to readily discriminate forms by noting differences in color, size, shape, etc., is so well integrated that the required behavior responses are largely unconscious. The heavy human dependency on the visual system in contrast to other aspects, such as smell, hearing, etc., is the result of an evolutionary process. The successful evolutionary survival of man, over more than 4 million years is, in part, a consequence of bipedalism (Washburn and Dolhinow, 1972; Washburn and Jay, 1968; Howells, 1973; Lestrel and Read, 1973; Lestrel, 1975; Conroy, 1990). This placed man into a unique eco-niche by totally freeing the upper limbs, resulting in a concomitant development of a comparatively large brain, and leading to an increasing use of the forward arrayed eyes (binocular vision being an important aspect within the primate Order). While a large, complex and developed brain is an asset, the use of binocular vision in conjunction with this large brain, was presumably important for the survival of early hominids. Thus, among the senses, vision in particular has played, and continues to play, an essential part in our adaptation to our external environment. While the initial details remain quite sketchy, it is not unreasonable to suggest that these evolutionary processes (such as the initiation of bipedalism, the subsequent development of a comparatively large brain, and binocular vision) played a critical role in the eventual development of human language. These steps led to the attainment of knowledge (and to the dispersion of knowledge to others in the group), initial steps which ultimately led to culture, science and technology.

Of the five senses (sight, touch, hearing, smell, and speech), we probably rely most heavily on vision, as already emphasized. However, all the senses come into play. The basic units of the sciences come from the utilization of the senses. Gerard (1961) has outlined these as: [1] *space* (metrics, *e.g.,* centimeters) derived from vision, touch, muscle sense and the vestibular system (balance organs of the ear), [2] *substance* (mass, *e.g.,* grams) from smell, taste, touch, muscle sense, and also vision, and [3] *time* (seconds) from primarily hearing. From these notions of space, substance and time, comes the notion of the existence of an *entity* or *object*. Such an entity or object, while potentially external with respect to the observer, is initially defined in terms of the

1

sensory system, which is internal to the observer. These entities comprise the basic sensory data leading, eventually, to descriptions of form: although, in the beginning this sense data was only treated in qualitative and often subjective terms. We will eventually return to the problem of how to capture, in numerical terms, the visual information that resides in all forms. This will be taken up subsequent chapters. The next section focuses on the fundamental importance of the visual system.

1.1.1. The Visual Process

Modern society bombards us every day with visual information of one type or another. Commercial advertising assails us with carefully designed forms of objects, often in bright colors to catch and hold our attention. All intended to convey a message that will convince us to readily part with our hard earned money and dash out and purchase the advertised product. However, this is only one example, and a rather trivial one at that (notwithstanding claims to the contrary of the advertising community), of the use of visual information. In actuality, we are dependent on our visual system in much more basic ways. In fact, our very survival depends on it. We depend overwhelmingly on vision and some 3 million sensory nervous fibers are involved in the transfer of optical information to the nervous system (Gerard, 1961).

From the moment we wake in the morning, we make use of our visual system. We detect whether it is daytime or nighttime. We identity objects that are around us and act accordingly to avoid obstacles, we operate on directions to locations we want to reach. We recognize people and places we deem important. Additionally, we are able to store and recall images of places where we have been, events we have experienced, and faces of persons we have encountered in the past. We also use the visual system in even more sophisticated ways such as in reading and writing in which visual, locomotor and language skills are all simultaneously required. All this implies that the structure of the human visual system is highly integrated in complex ways, with many of the functional processes involved in the reception and recognition of visual stimuli remaining incompletely known (Spoehr and Lehmkuhle, 1982; Zeki, 1993).

It is no accident that even the language that we use is influenced by space, substance, time considerations, and ultimately, vision. As Gerard (1961) has noted, the English language contains numerous examples of such metaphors as: we "apprehend" a meaning, refer to a "tangible suggestion" or a "weighty problem". We may say it "looks heavy", but almost never that "it feels green". The importance of the visual system influencing language is also clearly apparent from such terms as vision (*video*) itself meaning seeing. Vision is evident in the very word *evident*. No matter how intellectual we attempt to be, we cling to *insight*. A philosophy is a worldview (eine *Weltanschauung*), that is, something seen, while contemplation comes from *contemplor,* to gaze at. Consideration is from *considero,* to look at closely, etc.

The visual images that are perceived by the human eye are a combination of specific properties that are inherent in all forms; namely, consisting of size, shape, texture, color, movement, etc. We are able to function normally in the external world only by being able to rely on and process this visual information, the one exception being those who have lost their sight. Our visual system is so well developed that we are able to effortlessly classify and compare visual images and rapidly act on the information perceived. What is not so well developed has been our ability to measure or quantify this visual information for storage and future retrieval. In other words, in contrast to the human capability of rapidly identifying and classifying this visual information, the mathematical description of the content of these visual images has been extremely slow in forthcoming. The next two sections take up the issues of what exactly is meant by quantification and how this interacts with the elucidation of process an ultimate goal of biology.

1.1.2. A Dual View of the World

In this visual world, we are surrounded by forms of objects of one sort or another. In one sense one can consider these visually perceived forms as consisting of a duality. This duality is composed of two types of forms, those that are *artificial* or man-made, and those that are *natural*. The natural world includes the planets, geological formations, and biological organisms to name a few. Artificial forms will be considered below. How we approach, this duality has important practical as well as theoretical ramifications.

Enter most man-made structures in the world and you will generally be confronted by a dizzying array of verticals, horizontals, angulars, circles and regular curves, representing uniformity and regularity. Examples include large objects such as the Parthenon in Greece, the Twin Towers in New York, and objects of smaller scale such as automobiles, watches and even computer microchips. The list is endless.[1] These objects in a man-made world are often built to demanding standards, some requiring exacting precision (tolerances of ±0.005 of an inch). This is a requirement if the separate parts of, for example, an aircraft engine or a watch, are to fit correctly. Moreover, such precision in the manufacturing process is required if many copies of a complex item are to be reliably constructed. Thus, the emphasis in a man-made world is on uniformity, and exactness; that is, with little *variability* (which if it arises would be treated as error, and if great enough, in need of correction). The success of technological progress is undoubted, and we will briefly focus on the rise of scientific knowledge that produced it (Chapter 3). Nevertheless, it will also be demonstrated that this very success which has led to this modern technology, the rise of what have been called the exact or 'hard' sciences, is, in some ways, inappropriate for an understanding of the biological sciences and even more so of the social sciences. Reasons for this inappropriateness will be explored further in this chapter.

In contrast to artificial or man-made objects, naturally occurring phenomena, with very few exceptions (perhaps crystals, etc.) generally exhibit considerable 'irregularities' or variability when one views the individual members of a class of such objects.[2] As indicated above, this stands in direct opposition to the man-made world. Variability is of central importance in the natural world. Naturally occurring structures, whether large or microscopic, always exhibit variability (even identical twins will exhibit variability due to environmental influences). This is even the case of those forms that in some way display symmetry. Symmetry abounds; consider the form of plant leaves, clamshells, and even human anatomy. Nevertheless, because of variability, this

---

[1] It is, thus, not surprising that the architecture of Gaudi in Barcelona, Spain, stands out as a notable exception from the rule of regularity of artificial forms.

[2] However, even such naturally occurring objects tend to display variability, however small.

symmetry is never exact in the sense that the 'left' side is identical to the 'right' side. In biological organisms, the presence of this variability is an essential ingredient of evolution and can be considered an important facet of species survival.[3] Thus, this contrast between a man-made world (where one would like 'zero tolerance'; that is, no error or variability) and the natural world is profound and has a direct bearing on the use of quantification in the biological sciences. We will return to this issue in a moment, but first we need to examine why, in some ways, the physical sciences are simply not appropriate models for understanding biological processes.

There are a number of reasons for this state of affairs. These reasons can be summarized with three principles listed as: [1] *variability*, [2] *complexity,* and [3] *history*. The importance of variability has already been alluded to. The study of complexity, an emerging discipline in itself, refers to a basic aspect of all life forms. All organisms, whether single-celled or higher up the evolutionary ladder exhibit increases in complexity during their life cycles. The issue of complexity is briefly taken up in Chapter 5. The third principle, history, has to do with the fact that all organisms go through two types of developmental cycles. One is *ontogeny* or individual growth and the other is *phylogeny* or evolution. Both time-dependent processes lead to changes in the biological form.

These three principles distinguish the biological sciences from the physical sciences. It has been generally held since the start of the so-called scientific revolution in the 17th century, that variability (requiring the application of statistical theory), if its presence was acknowledged at all, played a decisively minor role in fields such as physics.[4] This is less the case currently, for example, quantum mechanics is now viewed in probabilistic terms. Nevertheless, the prevailing view, until recently, held that the physical phenomena observed in astronomy and physics was subject to readily predictable laws, which were unchanging with time. Although, it was recognized that the origin of stars as well as the origin of the universe, both implicitly implying change, such change was very slow, on the order of millions of years. Moreover, the orbital positions of planets could be predicted with a high level of reliability for many years into the future. Consequently, within this framework, the idea of evolution or change was not of major importance in physics.[5]

Two other aspects, associated with the successes of science since the 17th century, and which became the prevailing ideology at least until the middle of the 20th century, were the ideas of *mechanism* and *reductionism*. In brief, mechanism refers to the presence of regular order to the universe and everything in it, allowing prediction and mathematical laws. Thus, once the laws describing orbital motion were correctly deduced, process in physics and, especially, astronomy seemed to be immutable and unchangeable for all time. Reductionism refers to the ability to reduce all phenomena, including living organisms, to the same fundamental laws that comprise chemistry and

---

[3] To meet changing environmental conditions, species have to have the capability to adapt. This adaptation (based on mutation, the ultimate source of variation, and natural selection) is dependent on the gene pool, which must contain enough variability to insure survival in a constantly changing environmental landscape.
[4] The serious study of variability can be traced to the rise of statistics at the ending decades of the 18th and the beginnings of the 19th century (see Chapters 2 and 4).
[5] For a recent and novel view of the physical sciences from an evolutionary perspective, see the works of Sheldrake (1988; 1995).

physics. While at one level, living organisms are subject to biochemical laws, such knowledge has become of increasingly limited use in discerning the nature of those processes acting at the organismal level. As Mayr has cogently indicated:

> Living organisms form a hierarchy of ever more complex systems, from molecules, cells, and tissues through whole organisms, populations, and species. In each higher system, characteristics emerge that could not be predicted from knowledge of the components (Mayr, 1997:*xiii*).

The direct application of the laws of physics and chemistry to biology in a reductionist program, while attractive, has had only limited success. Such efforts should be largely abandoned because of the three principles: variability, complexity and history, preclude the direct application of such laws, except in specialized cases. The next section introduces the idea of quantification, which is a central theme of this volume.

1.2. THE ISSUE OF QUANTIFICATION

Why is measurement important? Three broad reasons can be provided. These are: [1] in the process of observing natural phenomena, the need to record its occurrence, [2] in identifying relationships and generalizing about the phenomena, and [3] in developing theories leading to prediction of new phenomena.

It can be said that without quantification there can be no mature science. This is largely true in the physical sciences. Nevertheless, progress in quantification in the biological sciences; that is, the 'softer' life sciences such as anthropology, biology, zoology, etc., has been considerably slower in contrast to the 'exact' or physical sciences (physics, chemistry, astronomy, etc.). As mentioned earlier, one reason for this state of affairs is that the objects of study in the biological sciences, at least on one level, are notably more variable and complex.[6] This has made it considerably more difficult to apply the quantitative methods that have proven so successful in the physical sciences (Bailey, 1967). Initially, with the rise of the natural sciences, the need for quantification of biological organisms was rarely contemplated, except for simple enumeration. Accordingly, it can be presumed that quantification was simply not considered particularly relevant to the issues being addressed. Therefore, most of the biological sciences were initially (and it can be argued that to a considerable extent, still are) largely descriptive. Fauna and flora, for example, were and continue to be described in terms of color, structural differences, behavioral considerations, adaptation to environments, and so on. These aspects determined similarity and differences between organisms and led to systems of classification (taxonomy). Another reason that may have initially acted to inhibit the use of numerical techniques, may be the fact that the human brain, concomitant with a highly well-developed visual system, provided descriptive information that served the biologist's purposes and answered the questions being then raised. It was only when the issues considered important changed; especially, as *process* in contrast to classification became more important, that it was gradually recognized that the descriptive approach, by itself, was not sufficient. The definition of

---

[6] This complexity arises because the subjects of study in the biological sciences are (or were at one time) living. This confers them with additional attributes not found in the materials studied in the physical sciences.

process as used here is intended to convey the following meaning. By process, what is broadly meant is the elucidation of those forces that shape the biological form during development, evolution, or due to functional or biomechanical constraints.

Any discussion of quantification in science must necessarily start with a basic discussion of the nature of measurement. That is, measurement must: [1] be an *operationally definable process*, [2] be *reproducible,* [3] have *validity,* and [4] be based on *aggregates* (Wilks, 1961). An operationally definable process means that the procedures involved must be defined in objective terms so that comparable results can be independently obtained by other investigators. Measurement may be as simple as counting or as complex as measuring with sensitive instruments, phenomena as elusive as the speed of light. The second basic requirement is that of reproducibility. That is, repeating the process of measurement to produce a reasonable agreement among different investigators. Unfortunately, the need for the reproducibility of measurements is often not cited adequately in the relevant scientific literature. This issue of reproducibility is briefly taken up in Chapter 2.

The third aspect, validity, is not to be confused with reproducibility. Validity is considerably more difficult to obtain. It does not refer to the replicability of repeated measurements, but rather to how close the numerical value of a measurement is to the 'true' value as either determined from an independent procedure or derivable from statistical considerations. Clearly one should strive for both high reproducibility and high validity. Unfortunately, the problem that arises is that such an independent procedure (or theoretical model or expectation) is the exception rather than the rule, especially in the biological sciences. Therefore, one has to rely not only on the experimental rigor of the measurement procedures used but also on logic to prevent contradictions or inconsistencies in the data and subsequent analysis. In addition, careful attention is required to insure that the measurements are derived from independent sources.

The last issue of quantification, aggregation, is concerned with what Wilks called 'aggregates or systems of measurements'. This is particularly important because you cannot base reliable conclusions on the study of a single individual object or single measure. This requires an aggregation in some fashion of the measurements from each individual member of a *sample* of objects drawn from the *population* of all objects. A primary and important purpose of such an endeavor is to arrive at some measure of the variability in the sample as well as estimating some statistical measures such as means, variances, and more sophisticated parameters. Moreover, for this process to be scientifically valid, it is necessary that the samples selected from the population be based on randomization principles. These three quantification principles are developed further within the framework of the scientific method (Chapter 2).

Finally, one may raise the question of why the need for measurement at all. Interest aside, how many undergraduate majors still choose the biological sciences in the hopes of avoiding the need to learn mathematics beyond the absolute basics. The reader may be assured that this is by no means a modern dilemma. As D'Arcy Thompson indicated:

> The introduction of mathematical concepts into natural science has seemed to many men no mere stumbling-block, but the very parting of the ways. (Thompson, 1942:11).

Additionally, description, in contrast to quantification, still forms a large part of the subject matter of the biological sciences today. Nevertheless, one reason for the use of mathematical descriptions in contrast to purely descriptive or verbal ones is that much greater clarity, objectivity and precision can be attained. Again, in D'Arcy Thompson's words:

> The mathematical definition of "form" has a quality of precision which was quite lacking in our earlier stage of mere description; it is expressed in few words or in still briefer symbols, and these words or symbols are so pregnant with meaning that thought itself is economized (Thompson, 1942:1026).

Finally, there is one more reason why a quantitative approach is to be preferred in the analysis of the biological form. The human visual system, while highly sensitive to movement and color, is considerably less sensitive to small structural details in a complex image or to subtle changes in the contour of an outline. When confronted with, for example, irregular outline data, unless the differences are pronounced, the visual system can be overwhelmed, with the result that fine distinctions can be missed. This is another justification for the use of quantitative descriptions of form. Moreover, today most published papers in the biological sciences will contain quantitative aspects (Chapter 2).

1.2.1. What is Morphometrics?

In a very basic sense counting and measuring are extensions of observation and induction, ancient methods that have been developed for the purpose of exactness and precision (Searles, 1956). Accordingly, all of us use morphometric principles on a daily basis, even if we are not consciously aware of the procedures involved. Our language even reflects morphometric descriptions. For example, when traveling and we ask how far is it to the next town we are asking a morphometric question. The answer of 25 kilometers represents a distance, in effect a morphometric variable of the type grounded in everyday experience. If we are attempting to repair the leg of a table, we will have to use a ruler and carefully measure the wood to be used to replace the broken leg. Again, morphometric principles come into play and lead to what might be called naïve morphometrics. The use of naïve here is intended in the philosophical sense (*e.g.,* a naïve realist). Clearly, morphometrics has ancient roots since the need for measurement goes far back in historical times, if not earlier (Chapter 3). The act of measurement then, is central to morphometrics, but as will be seen in this volume, modern morphometrics is much more than just measurement.

Although the etymology of the word 'morphometrics' seems to convey a straightforward meaning (Greek: *morph* = form, *metrikos* = measure), this is misleading for a number of reasons that will be made clearer subsequently. To begin with, the naïve morphometric examples given earlier generally represent solutions to practical everyday needs. While undoubtedly essential, and the very fabric of many occupations and professions, carpentry, engineering, etc., they tend to be ends in themselves. That is, once the measurements have been obtained, it becomes a relatively simple matter to construct the object (say a table leg) of interest. In other words, here mensuration is largely the means to an end. The validity of the measurements themselves is scarcely questioned as long as they are perceived to satisfactorily facilitate the desired result.

In contrast to such practical solutions, scientific endeavors cannot be so easily characterized because the measurements utilized may be complicated and often require sophisticated instrumentation and, more importantly, can have a direct influence on the outcome. This issue will be explored in some detail in chapter 4. Consequently, the application of morphometrics to biological data (the central focus of this volume dealing, after all, with biological morphometrics) turns out to be a considerably more challenging affair than might appear at first glance. Two reasons for this situation are: [1] none of the methods currently available are generalized enough to serve as a complete representation of form; and [2] of the lack of a formal unifying model underlying the measurement of form in terms of recognizable general factors that provide explanations for biological processes. With respect to these issues, little substantial progress has been made since the Zuckerman symposium on growth and form fifty years ago, which touched on some of these issues (Zuckerman, 1950).

The first volume to use the word morphometrics in the title was the pioneering work called *Multivariate Morphometrics* (Blackith and Reyment, 1971), which was followed with a considerably slimmer second edition (Reyment, *et al.*, 1984). These two volumes focused attention on morphometrics, although in rather narrow terms characteristic at that time.

Morphometrics is currently a dynamic field of study undergoing major changes. In particular, the last three decades have been instrumental in the development of new approaches, which have led to significant changes in the way morphometrics has been fundamentally viewed (Rohlf and Bookstein, 1990; Rohlf and Marcus, 1993; Marcus, *et al.*, 1996). Consequently, what has emerged are not only new, more sophisticated procedures, but an increased recognition of the need to model the morphological form as it really appears in two and three dimensions. Thus, since the 1970's, theoretical developments have made the comparatively new discipline of morphometrics into a flourishing, robust and expanding research area. Presumably, these developments will greatly accelerate in the next century.

Morphometrics consists of procedures, which facilitate the mapping[7] of the visual information of form into a mathematical (symbolic) representation (Read, 1990). Thus, morphometrics as viewed in this volume, consists of a considerably broadened definition that encompasses, at this moment, a number of separate and distinct approaches that are intended to deal with the numerical description of form. These approaches are briefly outlined below (Section 1.3) and represent the subject matter of Chapters 7 through 10. Ultimately, these diverse approaches will need to be merged into a coherent and unified model. Chapter 6 attempts to incorporate various aspects of an object into a formal model of form.

Finally, in parallel with the increasing number of morphometric methods aimed at numerically describing the biological form, there have also been developments in imaging, data acquisition and storage procedures. This has become a necessity for the

---

[7] Another word for mapping is *function* or *transformation* Given two *sets*, S and T, a function, *f*, is a mapping (rule) from S to T such that for any member of the *domain* set S, one can find a corresponding value in the *codomain* set T (the *range*). This mapping is generally *one-to-one* unless stated otherwise.

efficient collection, management and analysis of large data sets. It is of special importance for boundary outline methods (Chapter 9) and in the extraction of numerical information dealing with structure from complex images (Chapter 10), where one may need to repeatedly retrieve stored images. Nevertheless, *image acquisition* per-se, can be considered as a distinct and separate development from *numerical description,* the central focus of this volume. A number of authors have dealt with the issue of automatic 'image acquisition' (White *et al.,* 1988; Rohlf 1990; Jacobshagen, 1981; Glasbey and Horgan, 1995), and these may be profitably consulted by the interested reader.

1.2.2. From Morphology to Process

While quantification of the biological form must be viewed as one, if not the most essential step in the elucidation of *process* in the biological sciences, it is essential not to lose sight of the purpose of the scientific endeavor in the biological sciences. This is to provide explanations of process based on appropriate deterministically acquired facts. However, facts alone, either as descriptive data or subsequently as numerical data, while critically important to the research enterprise, in themselves, do not constitute explanations of process. As Read has noted:

> Directly or indirectly, form is central to our understanding of biological and genetic processes. The form mediates between internal genetic information and external environment; it is the means by which genetic information is evaluated and acted upon by natural selection (1990:417).

For example, it has been implicitly obvious for a long time now that the biological form not only changes its size but also its shape in rather complex, and at times, unpredictable ways during growth. In D'Arcy Thompson's words:

> And while growth is a somewhat vague word for a very complex matter ... It deserves to be studied in relation to form: whether it proceeds by simple increase in size without obvious alteration of form, or whether it so proceed as to bring about a gradual change in form ... (Thompson, 1942:15).

The biological form is subjected to numerous forces, some known and some potentially still unknown. Some of these forces are known to act concomitantly. These forces include long-term time-dependent processes such as evolution, as well as shorter-term ones occurring during growth and development. In addition, functional forces such as biomechanical loadings imposed on the individual during the life cycle; need to be considered (Oxnard, 1980).

Figure 1.1 (adapted from Giordano and Weir, 1985) is an attempt to illustrate the relationship of the measurement endeavor to the biological process, as a simple closed system. This closed system is composed of five distinct procedural steps, which are: [1] observation, [2] simplification, [3] analysis, [4] interpretation, and [5] verification. In brief, we start with a biological system from which we gather sufficient data to formulate a model. Data to be used in formulating the model must be: [1] sufficient and [2] simplified. Simplification is generally indicated; otherwise, it may become impossible to build the model.

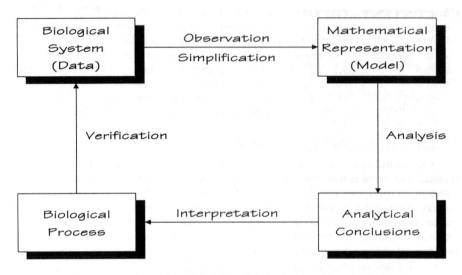

Figure 1.1. Modeling of biological processes.

The model is then analyzed and conclusions are drawn about the biological process being modeled. These conclusions lead to interpretations about the biological process. These interpretations, in turn, lead to changes, refinements and improvements in the implementation of the model. The validity of the model is then tested using new and/or independent data.

Ideally, this approach should guide the eventual development of tentative 'predictions' about the biological process. While such predictions, in all probability, will not have the exactitude of the physical sciences, their utility should not be discounted. In other words, numerical description should eventually lead to explanations of process. Nevertheless, as mentioned earlier, it is essential not to lose sight of the fact that morphometrics and the numerical data produced are not explanations of process (see Chapters 5 and 6 for more details). A discussion of some of these issues can also be found in Lestrel (1997).

Finally, if the focus is on time-dependent processes (*e.g.* growth), these require that changes in form be measured over time. Such measurements, if they are to precisely capture global changes of form, are not at all trivial and present a number of unresolved problems. These problems continue to persist, in part, due to a long-held but unwarranted assumption that the measurement procedures applied to biological organisms, are straightforward, easily invoked, and therefore requiring little consideration. That this is illusionary should become increasingly apparent as one proceeds through the chapters of this book. This volume then, is an attempt to provide an introductory background to these and other issues, as they relate to the description and analysis of morphological forms as viewed in the biological sciences.

1.3. CONTENTS OF THIS VOLUME

Currently, morphometrics, as viewed here, represents the combination of a number of distinct approaches dealing with the numerical description of form. Each utilizes data sets in unique ways and focuses on different aspects of the form. Moreover, these approaches have tended to be considered independently from each other because of a lack of a formal unifying model.

This book is divided into two parts. Part I, which is both introductory and theoretical in tone, is composed of chapters 1 through 6. The discussions here are an attempt to place morphometrics into a scientific framework starting with a brief description of the scientific method (Chapter 2).

One justification for approaching the study of modern morphometrics from a historical viewpoint is that by reviewing the activities and beliefs systems of ancient and medieval societies, can greatly assist in an understanding of the various developments that have led to modern science in general and to measurement in particular. Therefore, chapter 3 presents a brief historical treatment of the developments in science, especially where they impinge directly on the idea of quantification or measurement. The eventual development and increasing use of quantification in biology is covered in chapter 4. Part I ends with a discussion of complexity, leading to a tentative development of a formal, if heuristic, model of form (Chapters 5 and 6).

Part II, composed of chapters 7 through 10, focuses on the various morphometric methods currently employed by biological investigators. No attempt is made at completeness as the field is constantly changing, new methods are developed, and old ones are applied in new and novel ways. Four distinct strains of morphometrics can now be distinguished. The first one is *multivariate morphometrics,* which is typically applied to data sets composed of distances, angles and ratios (Chapter 7). *Coordinate morphometrics,* which focuses on deformations, including biorthogonal grids, finite elements and thin-plate splines (Chapter 8). B*oundary morphometrics,* is concerned with the boundary outline of the biological form, and involves methods such as median axis analysis, Fourier descriptors, eigenshape analysis, elliptical Fourier functions and wavelets (Chapter 9) and, *Structural morphometrics,* consisting of techniques such as Fourier transforms, coherent optical processing and, again, wavelets (Chapter 10). It is the author's hope that future research efforts will be devoted toward approaches that will provide for a more complete analysis of the morphological form, and especially how it changes in response to biological processes.

1.4. A NOTE TO THE READER

The purpose of this volume is to relate the subject matter consisting of the various strains of morphometrics into a unified and understandable fashion. This should allow the interested reader to see possible applications and encourage the utilization of these techniques in their own work. This book should not be construed as being, in any sense, complete in scope. That is not realistically possible given the explosive developments in the various disciplines concerned with morphometrics, rather it should be considered as 'work in progress'. It is intended, however, to reflect some of the current trends in the field as visualized by the author.

While technical terminology will be kept to a minimum, at times its use is unavoidable. Specialized scientific and philosophical terms are explained as they arise, particulars of which can be obtained from the references provided. A basic knowledge of mathematics is helpful, such as an elementary statistics course, matrix algebra and some exposure to the calculus. However, mathematics is not essential or stressed here for a basic understanding of the morphometric ideas presented. The material is presented with a minimum of mathematics so it should be understandable by a capable undergraduate or first-year graduate student in the biological sciences. Some equations are necessarily provided for the sake of completeness, but the details are largely omitted, since they are, generally, readily available elsewhere. The student is encouraged to initially skip those areas that may seem mathematically forbidding when first encountered. This will not affect an understanding of the subject matter. The emphasis is not on mathematics per-se, but rather on the utilization of the various techniques currently employed in the analysis of appropriate data sets.

Because of the author's background, initially in engineering and subsequently as a physical anthropologist, the emphasis will be primarily on the types of data encountered in the biological, medical and health sciences, but by no means limited to them. Further, to keep this volume within reasonable limits, space limitations preclude discussion of Procrustes analysis,[8] splines, fractals, fuzzy sets, neural networks, artificial intelligence (AI), chaos theory,[9] etc., all of which are also playing an increasing significant role in morphometrics. Thus, the techniques presented here are not the only ones in common use and no attempt at completeness is intended.

Finally, it might be useful to indicate what this book does not cover. While chapter 2 is a didactic treatment of research principles, it does not deal with statistics per-se. Statistical procedures form an independent although closely related topic to research methodology and books on that subject are readily available (Dunn, 1964; Afifi and Clark, 1984; Zolman, 1993). Although the subject matter of chapters 3, 4, 5 and 6 deals briefly with measurement, initially from a history of science framework and subsequently leading to a number of theoretical issues within what might be termed a 'science of measurement', no attempt is made at comprehensiveness.

Each chapter ends with a summary of key points, exercises to help illustrate the essential issues of the chapter, suggestions for student projects, and extensive references to assist the student with access to the available literature. Finally, a comment with respect to the exercises at the end of each chapter, these should increase understanding of the subject matter. Most of the questions are straightforward in nature; a few do not admit easy answers. These latter ones are intended to stimulate interest in, as well as gain an appreciation and awareness of, the numerous problems inherent in morphometrics.

---

[8] Issues dealing with Procrustes analysis can be found elsewhere (Siegel and Benson, 1982; Benson and Chapman, 1982; Chapman, 1990).

[9] Introductory materials dealing with, *e.g.*, fractals, fuzzy sets, neural networks, chaos theory, etc., are now widely available and easily obtainable.

KEY POINTS OF THE CHAPTER

This chapter introduced the concept of morphometrics as measurement within a scientific context. It touched upon the fundamental distinction between natural and artificial forms as seen by the human visual system. This was followed by a brief discussion of the inappropriateness of the uncritical application of mechanistic and reductionist notions to the biological sciences. The chapter then concluded with the need to understand that morphometrics is the first, critical, step toward the ultimate goal of explaining biological processes.

CHECK YOUR UNDERSTANDING

1. Why is morphometrics important in daily life? Cite some examples.

2. Given the examples in #1, some of which might be termed *naïve morphometrics*, contrast them with what you think would be a more rigorous scientific approach to morphometrics.

3. Describe the differences between artificial forms and biological forms. Why have artificial forms been easier to describe in numerical terms?

4. Make a list of English words that are influenced by visual considerations.

5. Can you think of a successful example of the application of reductionism in the biological sciences? Can you give an equivalent example for the mechanistic view?

6. Why is a quantitative explanation preferable to a descriptive one? In addition, when is it preferable?

7. Given the importance of process in biology, can you provide examples of some quantitative models, which have been successfully applied in the biological sciences? What characteristics do such successful models share?

REFERENCES CITED

Afifi, A. A. and Clark, V. (1984) *Computer-aided Multivariate Analysis*. New York: Van Nostrand Reinhold Co.

Bailey, N. T. J. (1967) *The Mathematical Approach to Biology and Medicine*. New York: John Wiley.

Benson, R. H. and Chapman, R. E. (1982) On the measurement of morphology and its change. *Paleobiol*. **8**:328-339.

Blackith, R. E. and Reyment, R. A. (1971) *Multivariate Morphometrics*. New York: Academic Press.

Chapman, R. E. (1990) Conventional Procrustes approaches. **In** *Proceedings of the Michigan Morphometric Workshop*. Rohlf, F. J. and Bookstein, F. L. (Eds.) University of Michigan Museum of Zoology Special Pub. No. 2.

Conroy, G. C. (1990) *Primate Evolution*. New York: W. W. Norton and Co.

Dunn, O. J. (1964) *Basic Statistics: A Primer for the Biomedical Sciences*. New York: John Wiley and Sons, Inc.

Gerard, R. W. (1961) Quantification in biology. **In** *Quantification, a History of the Meaning of Measurement in the Natural and Social Sciences*. Woolf, H. (Ed.) New York: Bobbs-Merrill Co., Inc.

Giordano, F. R. and Weir, M. D. (1985) *A first Course in Mathematical Modeling*. Monterey, California: Brooks/Cole.

Glasbey, C. A. and Horgan, G. W. (1995) *Image Analysis in the Biological Sciences*. New York: John Wiley and Sons.

Howells, W. (1973) *Evolution of the Genus Homo*. Reading, Massachusetts: Addison-Wesley Pub. Co.

Jacobshagen, B. (1981) The limits of conventional techniques in anthropometry and the potential for alternate approaches. *J. Hum. Evol.* **10**:633-637.

Lestrel, P. E. and Read, D. W. (1973) Hominid cranial capacity versus time: A regression approach. *J. Hum. Evol.* **2**:405-415.

Lestrel, P. E. (1975) Hominid brain size versus time: Revised regression estimates. *J. Hum. Evol.* **5**:207-212.

Lestrel, P. E. (1997) Introduction. **In** *Fourier Descriptors and their Applications in Biology*. Lestrel, P. E. (Ed.) Cambridge: Cambridge University Press.

Marcus, L. F., Corti, M., Loy, A., Naylor, G. J. P. and Slice, D. E. (1996) *Advances in Morphometrics*. Marcus, L. F. (Ed.) New York: Plenum Press.

Mayr, E. (1997) *This is Biology*. Cambridge, Mass.: Belknap Press

Oxnard, C. E. (1980) Introduction to the symposium: Analysis of form. Some problems underlying most studies of form. *Amer. Zool.* **20**:619-626.

Read, D. W. (1990) From multivariate to qualitative measurement: representation of shape. *Hum. Evol.* **5**:417-429

Reyment, R. A., Blackith, R. E. and Campbell, N. A. (1984) *Multivariate Morphometrics* (2nd Ed.) New York: Academic Press.

Rohlf, F. J. (1990) An overview of image processing and analysis techniques for morphometrics. **In** *Proceedings of the Michigan Morphometric Workshop*. Rohlf, F. J. and Bookstein, F. L. (Eds.) University of Michigan Museum of Zoology Special Pub. No. 2.

Rohlf, F. J. and Bookstein, F. L. (1990) (Eds.) *Proceedings of the Michigan Morphometrics Workshop*. University of Michigan Museum of Zoology Special Pub. No. 2.

Rohlf, F. J. and Marcus, L. F. (1993) A revolution in morphometrics. *Tree* **8**:129-132.

Searles, H. L. (1956) *Logic and Scientific Methods* (2nd Ed.) New York: Ronald Press Co.

Sheldrake, R. (1988) *The Presence of the Past*. Rochester, Vermont: Park Street Press.

Sheldrake, R. (1995) *A New Science of Life*. Rochester, Vermont: Park Street Press.

Siegel, A. F. and Benson, R. H. (1982) A robust comparison of biological shapes. *Biometrics* **38**:341-350.

Spoehr, K. T. and Lehmkuhle, S. W. (1982). *Visual Information Processing*. San Francisco: W. H. Freeman.

Thompson, D. W. (1915) Morphology and mathematics. *Trans. Roy. Soc. Edinburgh* **50**:857-895.

Thompson, D'Arcy W. (1942) *On Growth and Form*. New York: Dover Pub.

Washburn, S. L. and Jay, P. C. (Eds.) (1968) *Perspectives on Human Evolution*, Vol. 1. New York: Holt, Rinehart and Winston.

Washburn, S. L. and Dolhinow, P. (Eds.) (1972) *Perspectives on Human Evolution*, Vol. 2. New York: Holt, Rinehart and Winston.

White, R. J., Prentice, H. C. and Verwijst, T. (1988) Automated image acquisition and morphometric description. *Can. J. Bot.* **66**:450-459.

Wilks, S. S. (1961) Some aspects of quantification in science. **In** *Quantification, a History of the Meaning of Measurement in the Natural and Social Sciences*. Woolf, H. (Ed.) New York: Bobbs-Merrill Co., Inc.

Zeki, S. (1993) *A Vision of the Brain*. Oxford: Blackwell Scientific Pub.

Zolman, G. F. (1993) *Biostatistics*. Oxford: Oxford University Press.

Zuckerman, S. (1950) The pattern of change in size and shape. *Proc. Roy. Soc. (Lond.)* B. **137**:433-442.

2. AN INTRODUCTION TO RESEARCH METHODS

The unity of science consists alone in its method, not in its material.

The Grammar of Science (1911)
Karl Pearson (1857-1936)

In the course of coming into contact with empirical material, physicists have gradually learned how to pose a question properly. Now proper questioning often means that one is more than half the way towards solving the problem.

Physik und Philosophie (1958)
Werner Heisenberg (1901-1976)

2.1. INTRODUCTION

It is not an overstatement to say that the man's insatiable curiosity of about the physical environment has led to the development of science. Reasons for the incipient developments that eventually led to the rise of science probably lie within the psychological/physiological makeup of *Homo sapiens*. The human search for order in a chaotic and unsettling world led to the need to comprehend and control the environment.

Science, in its most basic sense, is concerned with the question of why things are the way they are. In other words, this means is that science is not only concerned with the questions of *what*, which leads to the observation and collection of facts; but also to the more fundamental questions of *how*, which focuses on explanation. This can be given concrete expression with an example. By making astronomical observations, one can discern that objects in the heavens tend to move in regular, if strange patterns. Consider, for example, planetary retrograde motion with the consequent slowing, moving backward and then forward again. The theory of how those paths were maintained was not at all apparent prior to the 16th century. Nevertheless, it led to a very sophisticated explanation as early as 150 AD. The then prevailing view, the Ptolemaic system, was one with the earth at the center and the sun revolving around it. This was to remain the accepted worldview for fourteen centuries, in spite of some obvious difficulties, which were recognized but not seriously followed up. Nevertheless, it was not until the Copernican revolution in 1543, which permanently placed the sun at the center of the solar system, that the earlier *Weltanschauung* or worldview was altered, or as some have proposed, shattered (Kuhn, 1957). Reasons for this slow change in worldview were not grounded so much in scientific explanation per-se but rather in the cultural milieu in which science was practiced and perceived. Some explanations for this state of affairs will be further examined in the next chapter. Moreover, it will be later argued that these worldview implications also have direct relevance on the development of quantitative models of biology (Chapter 5).

In this chapter, we first take up more the practical aspects of how the science endeavor proceeds and how it is viewed by its practitioners. That this scientific approach to the understanding the external world has been successful, is now undoubted. The application of such endeavors, over many centuries, has resulted in the soundest knowledge regarding our external world. Thus, there is no better way than to start with a detailed examination of what science and the scientific method are.

2.1.1. Definitions of Science

According to Olson, a broadly based definition of science can be given:

> ... as composed of a set of activities and habits of mind aimed at contributing to
> an organized, universally valid, and testable body of knowledge about phenomena
> (Olson, 1982:7).

Other more specific definitions of what 'science' is have tended to vary, but can be considered from two points of view. One such view is based on content and the other on process. A typical content definition might be: science is the accumulation of integrated knowledge. A process definition might read as: science is the act of measuring important variables, relating them and explaining these relationships in terms of theoretical principles (McGuigan, 1978). However, these two views are not independent of each other. It is probably more correct to say that the process view follows the content view. In different terms, this means that science is composed of *description*, which should be subsequently coupled with *explanation*, which in a mature science, for example physics, ultimately leads to *prediction*. A definition that connects the two above views is:

> Science is an interconnected series of conceptual schemes that have developed as
> a result of experimentation and observations (Conant, 1951:25).

To this definition, a caveat is added here in the sense that it should be understood that science is also a human enterprise. That is, scientific 'progress' often, if not always, tends to be the result of intellectual achievements by unusually gifted individuals and not by committees. Consequently, it is these human accomplishments that lead to paradigm shifts (Kuhn, 1970); that is, major developments that tend to change the *Weltanschauung*, or worldview. Eventual acceptance of such theoretical constructs (or theories), is based, in a large part, on the principle of verification leading to a general consensus view. Another definition that has been commonly used is:

> ... a systematically organized body of knowledge about the universe obtained by
> the scientific method (McGuigan, 1978:2)

What then is the scientific method? Clearly, it is much more than just solving problems or collecting facts. The scientific method has been defined as:

> ... a continual process of testing, modifying and developing ideas and theories in
> according with the dictates of the available evidence (Bailey, 1967:45).

At one level, it is simply an extension of ordinary common sense. At another level, it is extremely sophisticated procedure with very few individuals clearly comprehending what is involved. Consider the mathematical basis underlying the general theory of relativity, the unified field theory or string theory. Historically, the evolution of what is now called the scientific method can be seen as having progressed through a number of stages. Initially, there is a period that Bailey calls 'natural history' which is composed of rather naïve uncritical observations of phenomena. These observations are generally collected in an unsystematic fashion. They may be first hand or anecdotal, often collected simply because the observer had a particular interest. There is then an effort to organize these observations (facts) into a coherent framework. Finally, at attempt is made to provide an explanation for the phenomena in question. As Karl Pearson put it:

> The classification of facts, the recognition of their sequence and relative significance is the function of science, and the habit of forming a judgment upon these facts, unbiased by personal feeling, is characteristic of what may be termed the scientific frame of mind (Pearson, 1911:6)

Whether it is possible to pursue scientific endeavors in an objectively unbiased manner, remains open to interpretation, since subjectivity arises in the very act of selecting a research area to pursue. Nevertheless, good scientific work requires high ethical standards, an issue to which we will return to subsequently (Section 2.2.1). The scientific method then, consists of three distinguishable intellectual processes. These are [1] choosing facts, [2] developing a hypothesis to relate these facts and [3] testing the validity of the hypothesis. These steps are largely mental and require judgment and experience (Singer, 1959). Thus, a point not always understood or sufficiently appreciated is that:

> ... science cannot be learned from books, but only in contact with phenomena (Singer, 1959:266).

2.1.2. The Scientific Method

The general characteristic of all scientific endeavors, in contrast to other forms of knowledge, is that they are developed within a conceptual framework, which becomes articulated with a set of rules of procedure, called the scientific method. These methods may facilitate the construction of models, which, in turn, eventually lead to theory building. These theoretical constructs tend to provide ever-increasing knowledge about the external world. Acceptance of these theories is always subject to continual independent testing and confirmation by other practitioners. In time, results are either accepted or rejected by scientific specialists (falsifiability criteria[10]).

The scientific endeavor consists of a series of steps that start with the asking of questions that lead to a research problem. These research questions are generally based on the presence of some initial data and background knowledge. From the research questions, one develops working hypotheses. These hypotheses are then tested. Testing often involves the collecting of further data, which is analyzed and interpreted (refer to Fig. 1.1). From the results, one can then accept or deny the hypotheses (Resnik, 1998). It is from these tested hypotheses that theories eventually emerge. These two concepts, hypothesis and theory will be taken up further in Section 2.2.2.

The scientific method is composed of a number of elements, which provide a powerful method of inquiry. Careful attention to the specific set of procedures may (since there is never any guarantee) allow the elucidation of, heretofore, unknown

---

[10] Falsifiability criteria refer to the procedures used to verify or falsify a theoretical construct (theory). See especially the work of Sir Karl Popper in this regard (Popper, 1959; Corvi, 1997).

relationships. In this process, measurement, that is quantification, is, in some sense, generally required.

Although some of the sciences, such as physics, astronomy and perhaps chemistry, use more sophisticated measurement techniques than perhaps others such as biology, psychology, and sociology, all are dependent on quantification at some stage of the endeavor. Clearly, measurement, the subject matter of this volume, is therefore *central* to the scientific method. Further, by focusing on morphometrics, it will be demonstrated that quantification is now of equal importance in the biological sciences as it is in the physical sciences. The material presented below is, thus, universally applicable to all the sciences. The emphasis now shifts to a brief outline of the scientific method as it is commonly practiced, to provide a background and thereby allowing the placement of morphometrics into that context.

2.2. LIMITATIONS OF SCIENCE

In spite of the fact that scientific advances in nearly all fields have made unparalleled progress in the 19th and 20th centuries, it is distressing to see how little of the scientific approach permeates the thoughts and behavior of the average citizen. Science is either uncritically viewed in almost magical terms or alternatively blamed for all the ills seen in society. This is often a consequence of mistaking scientific endeavors for technological ones. There is a distinct difference between scientific research and technological achievements. Science should be viewed as a body of theoretical knowledge, and technology as the application of that knowledge. In practice, of course, considerable overlap is inevitable.

Science is limited to what can be observed. This critical point cannot be over emphasized. If it cannot be observed, in some way, by the senses, or with instrumentation, which is an extension of the senses, it cannot come under the scrutiny of the scientific approach. For example, questions such as the existence of angels (seemingly held to be true by a large percentage of Americans) can never be scientifically tested and fall outside the provenience of science. Such metaphysical issues must be left to the realm of philosophy and religion. These fields of endeavor are what McGuigan (1978) has designated as the metaphysical disciplines. These include literature, music, art, etc., disciplines that deal with issues of valuation (*i.e.,* requiring value judgments) and while no less important, lie outside the limits of science as defined here. However, this should not suggest that ethical behavior, requiring valuation by the practitioners of science is not an important consideration, far from it (Resnik, 1998). This issue is taken up in the next section.

2.2.1. Ethical Considerations

Perhaps the best approach to the subject of ethical responsibility is to start with an instructive example drawn from the early anthropological literature, a study that involves measurement. In this actual case, it is both the pre-conceptions present at the time and the presence of systematic bias that played critical roles. The example, taken from Gould (1996:109-112) examines a number of ethical issues.

In 1906, a Virginian physician, Robert Bennett Bean, published an article that compared the brain size of American blacks with whites. The hypothesis being tested by

Bean was to see if there were measurable physical criteria that differentiated the two human races. Thus, Bean set out to establish these differences on objective grounds. He did not use the more common measure at that time, cranial capacity, but rather the *corpus callosum*. Why he did not use cranial capacity will be explained subsequently. The *corpus callosum* is a structure that contains fibers connecting the right and left hemispheres. Moreover, it is composed of two parts, the anterior or frontal part is the *genu* and the posterior or rear part is the *splenium*. Given the assumptions of classic craniometry at the time, the *genu* was presumed to contain the higher mental functions while the *splenium* contained the sensory motor capabilities. Thus, Bean reasoned that the *corpus callosum* would be a good measurement. Consequently, he set out to compare the length of the *genu* against the *splenium*. His regression plot of *genu* vs. *splenium* yielded a good separation, although with overlap, of blacks and whites (Fig. 3.1 in Gould, 1996:110). Based on these results, Bean argued that whites because they displayed a larger *genu* had more brainpower or intelligence than blacks. If this has the familiar old refrain of racism, that should not be surprising and it clearly reflects the times at the turn of the century. Moreover, the same old tired arguments have not subsided; see Jensen (1969) and, more recently, Herrnstein and Murray (1994).

We now return to Bean, and the question that was raised earlier, why did he not use the actual size of the brain; that is, the more conventional cranial capacity measurement? The answer can be found in the addendum where black and white brains were found *not* to differ in overall size. This fact was conveniently dismissed and allowed Bean to conclude:

> So many factors enter into brain weight that it is questionable whether discussion of the subject is profitable here (quoted in Gould, 1996:111).

Nevertheless, Franklin P. Mall, Bean's mentor at John Hopkins, had become suspicious and re-measured the data. However, he added one critical element missing in Bean's method. By randomizing the data, he made sure that *he did not know the identity of the subjects*, black or white, as he was measuring them (what is now called a blind study). His regression results were considerably different from Bean's in that they now showed no differences between blacks and whites. The plot of *genu* vs. *splenium* showed an almost total overlap of blacks and whites (Fig. 3.2 in Gould, 1996:113). What had transpired here? In brief, Bean's black data was biased. There was a marked socio-economic discrepancy between whites and blacks at the outset of the study making the samples effectively non-comparable.

What lessons can one draw from these two studies? Apart of the fact that both blacks and whites samples did not reflect the society as a whole; that is, they did not represent randomly sampled groups of the total population, three conclusions can be singled out for comment. First, Hall understood the need to properly randomize the data prior to measurement, which Bean did not. Secondly, Bean's prejudices, in part determined by the social milieu, ruled his approach and analysis of the data. He ignored evidence that was contrary to expectation and drew only those conclusions, which were in accord with the racist tendencies he held. Third, quantification, however precise, does not inherently guarantee a correct conclusion, especially in this case were the experimental design is inherently flawed. Finally, it should be noted that while Bean's work stands out as a particularly noxious example of 'bad science', this is in no way intended to implicate the scholarly work done by other craniometricians at that time.

Although ethical principles (especially those which are 'morally acceptable') are always subjective and culturally bound; it is perhaps useful to provide some general guidelines that researchers should follow. Many scientific societies have devised codes of ethics to be followed. The following material is abstracted from the International Statistical Institute's declaration of professional ethics (Manly, 1992). This declaration is based on four principles: [1] obligations to society, [2] obligations to sponsors, [3] obligations to colleagues, and [4] obligations to subjects. Obligation to society dictates that research results are to be of high quality and disseminated widely in the public domain. Obligation to sponsors implies that any confidential information provided by funding organizations for example, will not be divulged without express permission. The researcher also has the responsibility to consider other appropriate alternative research designs and analyses if required, and to provide them with impartiality to the sponsor. Obligation to colleagues is especially important, as they are the primary reviewers of the research prior to publication. Thus, it is imperative that they are supplied with reliable information about the methods used in the preparation of the research. Adherence to collegial obligations also maintains and improves public opinion of the discipline and those professionals within it. Obligation to subjects is a major consideration when human subjects must be involved. Respect for privacy and the need for confidentiality must be always maintained. Informed consent must always be used. That is, the subject must be informed about the nature of the study and his or her role within it. To a lesser extend this applies to animal experiments, such as steps taken to minimize pain, etc. Generally, all bona fide institutions now require consent forms that need to be signed by the human participants in the study as well as approval is required (from human and animal protection committees), before current research can commence in the biological sciences.

2.2.2. Principle of Independence

A fundamental principle of scientific research is that the researcher can be considered as 'replaceable'. What this means is that those researchers, who independently repeat the observations or the experiment, should draw the same conclusions from the research. This approach is intended to produce consensus. This has also been defined as intersubjective validity or intersubjective verification (Grinell, 1992). It is this consensus that eventually leads to the adoption and acceptance of theories. As knowledge gradually increases in an area, facts and especially the relationship between facts becomes more understandable. This relationship has been called a hypothesis. A somewhat stronger construction is a theory. Here a test of validity tends to often be its predictive power. The question that subsequently arises is whether the hypothesis/theoretical construct can be viewed in sufficiently general terms (process of induction) to explain other facts. This consensus is subjected to continual testing until an exception to the rule arises, which leads to a new round of experiments and eventually, either rejection or refinement of the model. A good example of refinement is Einstein's versus Newton's view of physics. In this process of continual self-correction lies the strength of the scientific method. It is one of the great achievements of the scientific method that when such general hypotheses have been elucidated (becoming theories), they can have profound and far-reaching implications. One effect of the acceptance of these theories is that they act to substantially change the way we view the world around us. Consider Einstein's theory of relativity, or Darwin's theory of evolution, or Watson and Crick's discovery of the double helix.

2.3. SOME STATISTICAL CONSIDERATIONS

Whenever large amounts of data need to be reduced to a meaningful and manageable order, statistical techniques become essential (Searles, 1956). This requires that the biological investigator have a reasonable grasp of statistical principles. At the very minimum, these include measures of central tendency (means, medians, etc.), measures of dispersion or variability (variance, standard deviation, standard error of the mean), probability distributions (Gaussian or normal, skewed, bi-modal, etc.) and correlation. Basic statistical methods dealing with inference and description are now part of the exposure of most practitioners in the biological sciences. However, this exposure is often insufficient, witness the abuse and misapplication of one of the most commonly used statistical tests—Student's t-test.[11] Papers continue to be published containing tables with many variables, sometimes 20, 30 or more, the significance of each being individually tested with the t-test, in spite of the fact that this is an inappropriate application of the statistical test. The t-test was intended to test only the difference in means between *two* samples based on a *single* variable, not more than one. This is a caution that must be kept in mind with use of the t-test. From a statistical standpoint, the larger the number of individual t-tests, the greater the probability that some of these multiple comparisons will be statistically significant solely due to chance alone. Problems may also arise if these variables are in any way related or correlated. If one is concerned with *more* than two samples, then the appropriate statistic is an analysis of variance (ANOVA). If there are *more* than two variables involved, the proper test is a MANOVA or multivariate analysis of variance (Zolman, 1993). Although the increasing use of statistical techniques in the biological research literature is to be welcomed, there is still comparatively little utilization of multivariate statistical techniques, in contrast to univariate methods, in spite of the fact that they latter are often the more appropriate methods. The relevance of multivariate methods to morphometrics will be taken up again in Chapter 7.

According to Zolman (1993:3, see references therein), reviews of the literature have shown that close to half of published articles used incorrect statistical methods. Further, about 25% were found to be flawed because of confounded experimental designs as well as the misuse of statistical methods. In other cases, the sample sizes were so marginal that the possibility of finding a statistically significant effect was quite small. The lesson to be drawn here, and one that needs to be emphasized, is that along with the development of an adequate research design, the researcher must also make a reasonable effort to understand the statistical methods being applied and, perhaps, worry less about the computations involved. The latter issue is much less of concern now with the availability of statistical software packages for PCs requiring less reliance on central

---

[11] William S. Gosset (1876-1937) developed the t-test to handle the quality control of brewing beer using small samples. He published under the synonym of "Student".

university mainframe computers. Clearly, the researcher is strongly advised to seek the assistance of a professional statistician in this regard.

2.3.1. Bias toward the Use of Statistics

Finally, brief mention needs to be made of the issue of bias toward the use of statistics. Popular misconceptions about the use of statistics still abound. Consider statements like "figures don't lie, liars figure" or "statistics can prove anything" (Zolman, 1993). Others are "There are three kinds of lies: lies, dammed lies, and statistics"— attributed to Disraeli and "Round numbers are always false"— attributed to Samuel Johnson (Huff, 1954). It is, thus, unfortunate that even some biologists continue to maintain that a presumably well-designed biological study that provides *obvious* results does not need any statistics. While that view may represent a decreasing minority, there is still the tendency among practitioners in the biological sciences to readily accept results that present simple statistical tests (*e.g.,* the t-test) and tend to view with suspicion the utilization of multivariate statistical techniques such as the MANOVA, canonical analysis, principal components or discriminant functions, as an unjustified manipulation of research data.

2.3.2. Types of Research Studies

There are two major types of research studies commonly encountered in the biological sciences. These are the *observational* type of study and the *experimental* type of endeavor (Manly, 1992). In one sense one can consider the observational type of data to represent an example where the events are not controlled by the investigator, while in the experimental type of data, events are to some extent controlled by the researcher. An example of the observational approach might be the gathering of hospital data records to see if there is a correlation with a blood factor and the incidence of a disease. Another study might be the random allocation of subjects to two groups, one to receive an experimental drug and the other to receive a placebo or serve as a control group. Measurement, in one form or another, and statistical analysis need to be applied to either observational or experimental data.

In addition to observational and experimental type of studies, the investigator must also be aware of, so-called *between-group statistical* designs, and those, which are *within-group* ones. Different statistical procedures are involved with each (these are taken up in Chapter 7). In addition, the samples collected may be *independent* or *dependent* (related). Again, separate statistical tests are usually indicated. In the latter case, dependent samples, the observations are usually correlated which is a complication. An example is a growth study with repeated measures taken of the same subjects. All longitudinal studies are of this type. Specific tests such as the paired t-test and a repeated-means ANOVA and MANOVA have been developed to handle these data.

In sum, the scientific method requires an orderly and organized approach toward solving research problems, in effect, an experimental plan. While research methodology deals with the more practical procedures involved in carrying out a research project (Section 2.5) and eventual publication (Section 2.6), statistical considerations also play an important role. Thus, statistical issues as they impinge on morphometrics research, will be discussed as they arise (see *e.g.,* Chapter 7). We now turn toward a general approach in applying the scientific method to biological research. It involves three stages. These are: [1] the research plan or protocol, [2] the actual procedures to be used in carrying out the research, and [3] the writing up the results.

2.4. THE RESEARCH PLAN

An experimental plan or protocol is analogous to the use of blueprints to build a house. These blueprints translate ideas into practical working requirements so that the project can be completed. The *research protocol* should be considered as a prerequisite or initial step, from which the procedural endeavor (Section 2.5) is derived. In fact, there is considerable overlap between the two. However, the proposal does *not* contain critical elements present in the procedural endeavor such as the research results, discussion and conclusions, since these are not yet available. The distinguishing characteristic of the protocol is that it is intended as a document for departmental faculty review and approval as well others such as administrators, consultants, co-workers and technicians who will be involved in the research, and for funding purposes. Nevertheless, the research protocol is not a grant application although it contains material that is useful for the subsequent development of the grant application. In sum, a research protocol is intended to convey, in a convenient order, the elements of a particular research plan (Blandford, *et al.*, 1984a; 1984b). Table 2.1 is intended as a guide to the steps used to develop such a research protocol.

---

1. Statement of the Problem
2. Survey of the Literature
3. Research Objectives
4. Information Required
5. Data collection Procedures
6. Data analysis Procedures
7. Statistical techniques to be utilized
8. Significance of Proposed Research
9. Staff, facilities, equipment and supplies
10. Budget and Time Scheduling

Table 2.1. The research protocol.

2.4.1. Initial Steps

The preparation of the research protocol starts a the statement of the problem, survey of the literature, statement of the research objectives, data collection and analysis procedures, equipment needed, significance of and justification of the proposed research, budget and staff requirements as well as the need for required documents such as the animal and human protection forms, etc. These aspects are briefly listed below.

2.4.1.1. Statement of the problem

A concise description or summary of the research project, which states the problem to be addressed, presents hypotheses and the procedures to be used in attaining project goals. Hypotheses should be simple, concise, and consistent with the data. Reviewers of the research proposal need to be able to readily relate the objectives of the proposed research with the methodology to be utilized to achieve the stated goals.

2.4.1.2. Survey of the literature

New scientific breakthroughs do not appear de-novo; they are always based on previous knowledge. The quote "If I have been able to see further than others, it is because I have stood on the shoulders of giants" attributed to Issac Newton,[12] is the dictum by which all new research is initiated. That is, it starts with a careful and extensive survey of the available literature on the subject. It is from this library survey that scientific research problems can be identified and potential solutions provided. This section presents detailed background documentation for the proposed research project. It is important that the investigator conveys to reviewers a familiarity with previously published work.

2.4.1.3. Research objectives

Refers to the goals of the research project and should be set forth initially in broad terms, specifying what is to be accomplished. For example: "To determine whether a particular treatment procedure, X, results in a particular beneficial effect, Y". The following sections provide the details necessary to attain the project goals.

2.4.1.4. Information required

Section refers to materials that are now required in all research projects that use animals or are to be based on human data. These include human and animal protection forms and the required subject permission form. Still other forms may be required by a particular research institution or granting agency.

2.4.2. Collection and Data Analysis

Consists of the central part of the research plan and deals with data collection, data analysis, the statistical techniques to be used, and to convey what the significance of the proposed research is.

2.4.2.1. Data collection procedures

This is a description of the population to be studied. It includes characteristics such as the samples to be collected. These characteristics must be explicitly defined with sufficient detail such as sample sizes, specific ages, etc.

---

[12] While attributed to Newton, the quote actually originated with Bernard of Chartres in the 12th century (Grant, 1996).

2.4.2.2. Data analysis procedures

Here we now turn to how the project data is to be collected. What kinds of measurements are to be made? What kind of instrumentation is to be used? What kinds of variables are to be measured?

2.4.2.3. Statistical techniques to be utilized

What kind of statistical tests, such as measures of central tendency, of variation and possible multivariate statistical approaches, are to be used. Also to be considered are the kind of computer programs that may be involved. These need be identified and discussed.

2.4.2.4. Significance of the proposed research

Why is this research project important? This section is of particular concern for the reviewer because in it the investigator must provide a justification for the significance of the study and why it should merit approval and funding.

2.4.3. Some Other Requirements

Other requirements include equipment and supplies to be utilized, also involved are budgeting considerations.

2.4.3.1. Staff, facilities, equipment and supplies

Depending on the size of the research project, its duration, and projected costs, considerable thought has to go into such issues as staffing, facilities and especially equipment, which may be a substantial part of the budget if it has to be purchased.

2.4.3.2. Budget and time scheduling

Section refers to budgetary requirements as well as a proposed timetable for completion of the project. A realistic estimate of both budget and time is required. These are necessary to justify the expenditures of the potential granting agency. Finally, as there is considerable overlap with the proposal and the actual procedures involved in carrying out a research project, those elements, which are held in common by both, will be further elaborated in the next section (2.5).

2.5. THE PROCEDURAL ENDEAVOR

The scientific method represents the cornerstone of all research, regardless of the field or discipline. Scientific problem solving is never haphazard; it is the result of an orderly process consisting of specific steps (McGuigan, 1978; Westmeyer, 1994; Thomas and Nelson, 1996). The sequence of the actual steps involved in the procedural endeavor is outlined as a checklist in Table 2.2.

The steps displayed in Table 2.2 however, cannot be viewed, as a set of specific procedures that need only be followed and results will somehow be guaranteed. The scientific method is not just a series of actions to be slavishly followed but rather what Thomas and Nelson (1996) describe as "as a disciplined inquiry" in which flexibility is essential. What this entails is that legitimate scientific knowledge is the result of experience, careful application of technique, and most important, hard thinking. More often than not, a little imagination, or creativity, is also required. On the other hand, the lack of a research plan, poorly thought out experimental procedures, or insufficient sample sizes, represent circumstances that largely preclude the possibility of meaningful results.

---

1. Survey of the Literature
2. Statement of the Problem
3. Formulating Hypotheses
4. Selection of Research Design
5. Identification of Variables
6. Determination of Samples
7. Methodological Procedures
8. Evaluation of research results
9. Relationship of Results to Hypotheses
10. Generalization of the Findings

Table 2.2. Plan of research procedures.

Thus for success, it is imperative that careful systematic planning is required at the *initiation* of any research study. It is important to identify potential problems that need to be resolved before data collection is started. Often, the assistance of a bio-statistician can be helpful at the inception of the research process. These steps will be briefly outlined in the material below. The process starts with the recognition that a problem or controversy exists.

2.5.1. From the Literature Search to Hypotheses

The procedures generally start with a literature search to facilitate problem recognition, which leads to hypothesis formulation. These procedures are outlined in the next three sections.

2.5.1.1. Survey of the literature

As already mentioned in the protocol, one of the most crucial steps in the research enterprise is the survey of the literature. This step cannot be overestimated. It is regrettable to see how many practitioners in one field have no idea of parallel developments going on in another area, work that is directly relevant to their own research project. Another often held belief, which is erroneous, is that the older published data is somehow no longer relevant to current work.

The starting point toward the identification of a problem requiring solution starts with a review of the literature. This review of the literature entails utilizing *all* resources that are available. These include the journal literature, a primary source, as well as books, masters and doctoral theses, reviews, even on occasion, the more popular magazines such as Scientific American, American Scientist, etc. A few suggestions are in order here that may be found useful in the initiation of such a library search. Currently, this often leads to a search on the Internet where a number of search engines can be profitably utilized. The assistance of reference librarians in setting up search criteria may also be beneficial.

Once a relevant reference, say a journal article, has been identified, one then may have to physically locate it. Locating the article usually entails searching by the year of the bound journal. With the journal article in question in hand, the next step is to assess the cited references within the article. In this way, other relevant articles can be quickly identified and located. In addition, it is a good idea to scan the table of contents of the bound journal in question. This table of contents scan should cover a number of volume years, both forward and backward. The reason for this approach is that relevant further articles can be identified, which are often missed by the search engines.

2.5.1.2. Statement of the problem

This leads to a statement of the problem, defining it, and setting its limits. Often the problem can be stated as a question to be answered. Given that in most established areas of science, considerable research already exists, this makes the development of novel, new and significant knowledge, a challenging affair requiring creativity. It is not an exaggeration to say that solving significant problems requires considerable imagination as well as intelligence (or 'genius', however defined). Some of the ways in which problem areas can be recognized have been identified by McGuigan (1978). These are: [1] the presence of controversy in which contradictory results are present, [2] the absence of data in a specific area and [3] the presence of 'isolated' facts that do not fit with conventional theory.

Often, these problems can be better recognized in terms of relationships. For example, running a particular experiment leads to a specific result. That is, the result is related to the experimental procedure. However, considerably more is needed. One has to define what is to be measured. In a technical sense, this often involves the identification of independent and dependent variables. In physics, these may be directly related by a simple function such as $F = ma$; that is, the independent variables mass (m) and acceleration (a) influence the dependent variable force (F) in a multiplicative way. Another similar well-known function is $E = mc^2$. Here mass (m) times the speed of light squared (c^2), the independent variables, equal energy (E) the dependent variable. This relationship, while proposed by Einstein at the beginning of the 20th century, took many decades before it could be experimentally verified.

Such relationships of course, can be quite involved requiring hundreds of variables. In addition, there are other ways to study relationships. In a clinical context for example, one might be interested in changes in shape of the cranial base of hydrocephalic individuals (an abnormality consisting of premature closure of the cranial bone sutures) compared to normal individuals (those that do not exhibit the abnormality). In this example, one might think of the measurements of the hydrocephalic cranial base form as dependent, or treatment variables, while those of the normal individuals act as independent or control variables.

It is from these statements of relationships between variables that hypotheses are developed. In fact, one definition of a hypothesis is as follows: it is a statement of a potential relationship between two or more variables. Eventually, with repeated independent testing leading to confirmation, these hypotheses can evolve into generalizations, principles or theories. The older use of the word 'laws' is now to be avoided. From a consideration of these hypotheses (of which there may be more than one), flow suggestions of how the variables might be related, and thereby shed light on the problem under consideration. Once the variables have been determined, the experimental design can proceed. The next sections attempt to formalize this process called the scientific method.

2.5.1.3. Formulating hypotheses

The hypothesis is simply a statement of relationships, albeit often couched in formalistic statistical terms, of a result that is to be expected. Thus, it is not likely that one can carry out a research study without clear objectives. They may be very vague at the outset, or very specific. There may be one major hypothesis or many. At the most informal level, they may represent initial intuitive guesses, which are subsequently articulated as simple expressions of possible relationships. More often than not, imagination and serendipity are involved. However, hypotheses do not arise *de novo*. They need to be seen in the context of previously accepted observations and theory. These prior assumptions determine and limit what can be done. Thus, pre-existing theory or observation can mark the starting point for the development of new relationships (Grinell, 1992). Eventually, these relationships need to be formally re-cast into hypotheses.

The foundations that lead to hypotheses are generally based on [1] the outcome of an experiment, [2] on observations, [3] on previous studies, or [4] on theoretical principles. Often it is a combination of all of these. Notice that all of these procedures require that the hypotheses are *testable* in some way. Untestable hypotheses, which unfortunately abound, are unsolvable (for reasons that will be made clear) and hence, by definition, inadmissible in science. To give some examples: [1] human intelligence is superior to that of other mammals or [2] although earthquakes seem to appear as random events, they are not caused by forces of nature but represent God's handiwork. Such hypotheses cannot be reasonably tested because the first one is stated in too vague terms and the second one lies outside the realm of science. Moreover, with the latter one, the difficulty is not that such a hypothesis may be true or false on logical grounds, but rather that it cannot be refuted empirically. Now we turn to the more practical issues of conducting a scientific endeavor starting with the setting up of hypotheses.

Why go to the trouble of stating hypotheses in a formal manner? One reason is that it helps clarify the thought processes that underlie the purposes of collecting data or conducting an experiment. It also forces the investigator to focus on the variables that need to be selected. Hypotheses need to be stated in the form of synthetic propositions (Ayer, 1952). This arises from the fact that one needs to test the agreement of a hypothesis (which may be in error) against the empirical world. How then do you go about setting up a hypothesis? Bertrand Russell one of the most influential philosophers of the 20th century proposed the logical structure of a hypothesis in terms of an '*If_*, *then_*' construction. In other words, if certain conditions hold, then certain other conditions should also hold. According to McGuigan (1978), although most hypotheses are not stated in the explicit form suggested by Russell, they can often be re-cast into the 'if *a* then *b*' form. The following example is intended to illustrate this point:

> The purpose of this study is to compare the cranial growth of hydrocephalic subjects (exhibiting premature closure of cranial sutures) who have been treated with a shunt (a procedure to alleviate the increased intercranial pressure) with the cranial growth in normal individuals (Lestrel, *et al.*, 1996)

The implied hypothesis in can be restated as follows. *If* the shunt procedure is successful *then*, growth in the hydrocephalic subjects should approach normality. There are two measurable events here: The growth of the shunt-treated hydrocephalic subjects and the growth of the normal individuals (acting as controls). These two aspects need to be re-defined in terms of an identical set of variables to be applied to each group before the research can begin. This type of hypothesis construction can be defined as a *qualitative* hypothesis that it is based on the logical construct of 'if *a* then *b*'. Qualitative refers to the use of the English linguistic terminology instead of mathematical symbols (see next paragraph).

A second approach used to set up a *quantitative* hypothesis is derived from mathematics. In this case, we set up the hypothesis in terms of a simple function. That is, $Y = f(X)$, that is, Y is a function of X. Such a formulation clearly shows that the two variables are related. Another way to view the relationship is as Y is the dependent variable and X is the independent one. Although the hypothesis is now stated in mathematical terms, the notion of 'if *a* then *b*' still obtains. In this case, we would say, "*If* (and only if) X is this value, *then* Y is that value" (McGuigan, 1978).

Finally, three cautions are in order. The first one has to do with causality. That is, there is no implication here that *a causes b*. The second caution re-emphasizes that there is no certainty involved here. It is important to remember that although the logical construct "if *a* then *b*" can only be true or false, the actual hypothesis, based on empirical data, must still be couched in probability terms. That is, the hypothesis is probably true or probably false. The third caution deals with the issue of accidental occurrence. This refers to the case were 'if *a* then *b*' arises by accident. This refers to what in biology can be called a spurious correlation. This implies that for the relationship 'if *a* then *b*' to hold, there has to be a real and not 'accidental' relationship between the variables. For example, there could be another factor, say '*c*' which is actually responsible for presumed relationship (or correlation) of *a* and *b*. Finally, this section concludes with some of the criteria proposed by McGuigan (but modified here) that define 'good' hypotheses. These are listed in Table 2.3.

As part of setting up the hypothesis in formal terms, as well as the basis of quantification, we need to identify the relevant variables that we plan to measure. While the hypothesis represents a general statement of the relationship between variables to be measured, it is often not in specific enough terms. The choice of what to measure is not a trivial exercise by any means. The selection of variables tends to play a significant part in the results of a scientific investigation. In addition, this subject tends to be under emphasized in spite of its importance. There is the assumption, widespread in much of the sciences, that measurement is cut and dried, needing little comment.

1. Should be plausible, have clarity and be parsimonious (Ockham's Razor); that is, have logical simplicity,
2. Should go some way toward answering the initially stated problem,
3. Must be subject to testing and the testing must be repeatable,
4. Should be expressed in quantitative form and,
5. Should have a wide-reaching consequences; that is, it should be general in scope.

Table 2.3. Attributes of hypotheses.

One of the primary purposes of this volume to demonstrate that not only is this assumption a fallacy, but to suggest that the neglect of the measurement issue has hindered progress in the biological sciences. Thus, faced with the choice of what variables to select we come closer to one of the central issues of this volume. The question of variable selection will play an increasingly major role in the latter chapters of this book. We now turn to some of the more technical details of development of a research study, starting with the selection of samples.

2.5.2. The Research Design

These sections contain the crux of all research projects. They start with the research design, which will require identification of relevant variables to be measured on specific samples to be collected. Intimately involved are the methodological procedures to be utilized.

2.5.2.1. Selection of Research Design

This is where research project begins to take form. The research design is intended to relate the hypotheses to be tested with the judicious choice of collected samples and the selection of relevant variables. Use only controlled and repeatable experiments to study the phenomena under consideration. Collect data using the most reliable and sensitive instruments available, record, and save all measurements. Evaluate all errors due to instrumentation. More details will be found in Sections 2.5.2.2 and 2.5.2.3.

2.5.2.2. Identification of Variables

In the development of hypotheses, emphasis was placed on the need to think of hypotheses as relationships between variables. The choice of what variables to utilize in the research design is a critically important step. An inappropriate choice can lead to an inability to detect a difference, *e.g.,* between treatment and control groups, and at worst lead to false conclusions. Thus, careful consideration is indicated to insure that sufficient sensitivity is present to detect statistically significant changes in the measured variables in question.

2.5.2.3. Determination of Samples

While one often focuses on individual specimens or objects, one at a time, during the course of a scientific study, this is, in a sense, misleading. It is important to realize that science is based on a collection, or aggregate, of data values derived from such individually studied objects. That is, science is the study of *classes* rather than individual objects (Grinell, 1992). Consequently, to be able to test hypotheses, one needs to obtain representative samples of objects drawn from a population. Thus, the identification of these aggregates, or groups, becomes a crucial part of the process.

In any research enterprise, two primary steps are involved in the process. The first step involves the selection and number of subjects or specimens that are to comprise the sample and the second one deals with the techniques or methods that are to be applied to measure specific aspects of the sample. This second step is somewhat independent of the first and should not be confused with it (Section 2.5.2.4.)

The data for a research study can be obtained in, generally, three ways: [1] from already existing records, [2] by sampling to obtain new records and [3] by carrying out an experiment. Again, it is important not to confuse the two distinct elements here which are [1] the sample(s) to be utilized and [2] the procedures or measurements to be gathered from the sample(s). Implicit in the choice of what samples to collect is the experimental design. The first question that needs to be answered is what groups (experimental, control, *e.g.*) are to be compared and how many. The second question refers to how the objects (subjects or specimens) within a group are to be randomly allocated.

For example, one might choose a sample on which measurements are to be taken which is then to be compared with a control sample. The choice of which sample(s) is very important, since it is generally not realistically possible to use all available existing records or sample all subjects in a population unless it is very small. This immediately leads to the question of how to select such a random sample. It is of crucial importance that the sample is representative of the population from which it is drawn. Possibly, a more complicated design may be necessary requiring stratification or cluster sampling. Many sampling designs are now available, such as nested, randomized block or factorial designs, depending on the nature of the data (Manly, 1992; Zolman, 1993). Reference should be made to the numerous statistics books available on the subject. Although data collection is a seemingly straightforward process, there are pitfalls. One major issue that needs to be considered is whether the samples are large enough to provide reliable statistical estimates.

It has increasingly become a requirement of grant proposals that the size of the samples to be collected is estimated beforehand. While this can be difficult, especially since the standard deviations of the observations are often unknown in the absence of a pilot study, estimates can be computed. Thus, it is a useful and instructive exercise and any assessment of the sample size, even if subject to uncertainty, is better than no assessment at all (Manly, 1992). The calculation of the sample sizes necessary to detect the desired effect is called a *power analysis*. Reference should be made to Marks (1982) and to Runyon and Haber (1980) for their discussions of the concept of power and its relationship to type I and type II errors. The reader is also referred to Manly (1992) and Marks (1982) for examples of how to calculate estimated sample sizes. Finally, other statistical methods such as bootstrapping may be useful for sample size estimations.

2.5.2.4. Methodological Procedures

Once the samples have been identified and the measurements chosen, one can proceed to the second step; namely, applying techniques or methodologies to the data that will eventually promise results. This refers to the experimental procedures to be used. These techniques can range from the simplest such as counting or measurement with a ruler, to the use of very sophisticated machinery such as scintillation counters, electrophoretic techniques, as well as host of other procedures commonly in use in the biological sciences. The choice of measurement (broadly defined here) will have a major bearing on the results, so considerable thought should be expended on the type of measurements to be applied to the problem to be investigated. This includes what measurements to consider and how many. It is essential that whatever collection procedures be implemented, that they are carefully followed to keep experimental errors to a minimum. In addition, the possibility that bias may be unconsciously influencing the choice of measurements must always be considered. It is scarcely necessary to indicate that exactly the same measurements must be carried out for the treatment and control samples. Developing a protocol as outlined in Section 2.4 can be very helpful and it is highly recommended (Blandford, *et al.*, 1984a; 1984b). The protocol (Table 2.1) is often a useful precursor of what will eventually become a manuscript for publication (Table 2.4). It is always advisable to pre-test techniques prior to the actual experimental procedure, to be sure that all the investigators and technicians have mastered the numerous processes involved. This may disclose imperfections in the procedures that need to be corrected. Sometimes more elaborate checking of procedures is needed, which requires that a pilot study be instituted. It is especially important to keep the experimental error within acceptable limits. A procedure must also be instituted that allows for the computation of measurement error, which is taken up now.

While much of the theory of the propagation of errors properly belongs to statistics and the reader is directed to that literature, a few comments may be in order. To begin with, every measurement that is observable, no matter how precise, is still subject to measuring errors. This is often described with a range ($\pm$) within which there is a high probability that the correct or 'true' value will lie. One way to establish these limits is to have repeated trials of the measurement in question. Historically, the formal recognition of experimental errors in science can be traced to the popularity of games of chance in the 1650's in France. As considerable sums were involved, this led to attempts to compute gambling chances. Influential gamblers sought the assistance of noted mathematicians of the time like Pascal and later de Moivre. With mathematicians taking an increasing interest in these issues, the classical theory of probability was born (Parratt, 1961). Around 1738, de Moivre published what was the first suggestion of the normal probability distribution, later made famous through the work of Gauss. What became the central limit theorem was first applied by both Gauss and Laplace to the analysis of measurement errors in astronomical observations. However, after Laplace's work of 1812, formal probability theory lay largely dormant. With the exception of actuarial work and in certain areas such as biological measurement (chapter 4), probability did not effectively reappear until the end of the 19th century. This new approach eventually led (in 1927) to Heisenberg's now well-known uncertainty principle (Parratt, 1961).

Since randomness (stochastic variation) is an inevitable consequence of the application of the scientific method to sense data, we need ways to describe and precisely measure its presence. There is no such thing as an exact measurement. All measurements are subject to errors, however small. Errors associated with measurement can usually be identified as composed of two distinct types of variation. These are: [1] due to random or statistical errors and [2] systematic errors (Meyer, 1975; Healy, 1989). Statistical errors, which are generally random, tend to be values, which are equally high or low around a mean or true value. Thus, these statistical errors are 'regular' in the sense that we can predict the average or mean and their dispersion about the mean. One procedure for computing this statistical error is a replicability analysis; that is, carrying out repeated measurements. However, there is a caveat here. When measurements are repeated multiple times, it may occur that one value seems rather large. Unless this value is clearly anomalous, there may not be any justification in rejecting it. Large fluctuations do occasionally occur and are not, in themselves, abnormal (Meyer, 1975).

Systematic errors, on the other hand, often tend to be constant. An example is a calibration error in an instrument resulting in a too high reading. Another source is computation error due to mistakes or outright blunders. Careful consideration of every step in the experimental procedures is required to find and correct such systematic errors.

2.5.3. Research results

These are also three critically important sections as they are concerned with the results, their generalization and how those results are related to the initial hypotheses.

2.5.3.1. Evaluation of research results

Once the experiment has been run, or the relevant measurements taken, one is ready to analyze the results. The evaluation of the data requires some manner in which the reliability of the results can be analyzed. This procedure is independent from the issue of measurement error discussed above. In other words, are the results meaningful or 'real' in some sense. This step, in most cases, will eventually require as part of the process, the application of mathematical and/or statistical principles. In the absence of such statistical testing, it difficult to draw significant conclusions about differences between groups. For example, to consider whether the differences observed between treatment and control groups are meaningful. Moreover, it is now difficult to get research results published in quality journals without the careful consideration and application of statistical tests. It may be of considerable help to consult a bio-statistician *before* this step is reached to insure that the experimental design is appropriate. Questions that may arise, for example how to handle missing data, etc., will need to be addressed.

2.5.3.2. Relationship of results to the hypotheses

It is at this step that the investigator provides the evidence, as demonstrated from an analysis of the results, which will either support or reject the hypotheses set forth earlier.

2.5.3.3. Generalization of findings

At the stage, the researcher compares the results with those of others and, perhaps, attempts to integrate the results into a model (Chapter 5). Presumably, the findings are then prepared for publication (section 2.6), which will allow scrutiny of the results by the scientific community at large.

Finally, it then becomes the responsibility of future researchers in the scientific community to replicate the results. Researchers are then able to contribute to the development or substantiation of a theoretical framework (or theory) that has general applicability. In this way, our knowledge of the world around us is increased and at other times, corrected.

2.6. THE DOCUMENTATION ENDEAVOR

This section may seem self-evident and thus redundant; however, years of teaching have led me to the conclusion that students, especially at the beginning of their careers are not sufficiently familiar with the procedures involved in scientific writing. Clearly, the most brilliant achievement, worthy of the Nobel Prize, will remain unknown unless it is written down and eventually published. Publication continues to be the lifeblood of science. It is with publication that the academic community gets an opportunity to evaluate the validity of the research. Ideally, this does not mean all research should be published, only quality research. Given the large number of papers that already exist and continue to be published, this would seem a desirable goal. Unfortunately, the pressure to publish has resulted in a rough doubling of papers per decade. This is approximately 7% a year with over 7 million items available in 1976. Moreover, the average article is cited less than one time a year and around 50% are never referred to again (Brunette, 1996).

---

1. Title
2. Author(s)
3. Abstract
4. Introduction
5. Materials
6. Methods
7. Results
8. Discussion
9. Conclusions
10. References Cited
11. Appendices, etc.

Table 2.4. Writing up the research results.

The better and more prestigious journals are refereed ones, meaning that all potential manuscripts sent to them for publication are reviewed by peers. The peer review process is, however, by no means perfect. Some journals have considerably higher standards of what is acceptable quality work than others. Moreover, the peer review process in no way removes the researcher's responsibility for his submitted work.

It is still the ethical responsibility of the author(s) to submit manuscripts that are composed of high quality work with a minimum of errors.

Every manuscript submitted to a journal should initially describe what the problem is, why the research was conducted, how it contributes to our knowledge and to what areas the results found are relevant. Moreover, these elements should be presented in complete enough terms so that readers can judge the validity of the work and its conclusions. This is a challenging endeavor if a quality paper is to be produced; thus, writing up the results of research is never easy. It takes much practice to write well-organized papers that communicate ideas concisely and clearly.

Conveniently, Section 2.5 discussed earlier, contains, in considerable part, the information needed to write up the results for publication. Every major refereed publication publishes a *Guide to Authors*, which will list specific requirements such as the form of the entries in the references cited section. These guides are generally published once a year and can be found by perusing the journal in question. The steps involved in writing up the results for publication are discussed in the following sections and outlined in Table 2.4.

2.6.1. Introductory material

This includes basic items such as the title, author(s), abstract, if required, etc., as well as the introduction to the actual research.

2.6.1.1. Title

The title of the paper should be given considerable thought. If possible, the subject matter should be readily detectable in the title. Its length should be kept as short as reasonably possible. It should clearly inform the potentially interested reader of the subject matter to be covered. The title should be unique. Phrases such as "A study of ..." should be avoided since they are generally redundant.

2.6.1.2. Author(s)

Underneath the title of the manuscript are placed the author(s) names. Along with the names of all the authors involved, their institutional addresses need also to be provided. The recent trend toward multiple authors is, to some extent, regrettable, as it becomes less clear who contributed what to the paper. Authors chosen for inclusion should always be based on appreciable contributions, however they are defined. Minor contributors, such as those whose sole task consisted of providing the data or lab facilities, can be properly acknowledged in a footnote. The title, the authors and their institutional affiliations are usually placed on the first page of the manuscript to be submitted for publication. This first page, also called a cover sheet, should also include a set of 3-6 key words for indexing purposes as well as a shortened title to act as a running head to be placed on each final published page.

2.6.1.3. Abstract

This is a summary of the paper and often represents the only information that a potential reader will scan. If the manuscript is accepted for publication, the abstract will often be subsequently published in what are abstracting services such as the Medline, Dental Abstracts, Psychological Abstracts, etc. It usually consists of 200-250 words, written on a separate sheet of paper and placed at the beginning of the manuscript after the title page. It is often the last thing written, generally after the manuscript has been completed. In addition to justifying why the article is significant, it should also briefly discuss the purpose, samples used, methods, results and conclusions. The results are most important as well as the statistical procedures used. If the abstract is, in some way, incomplete or methodological aspects are improperly detailed, or it is poorly written, there is a high probability that the published paper will be ignored by the research community.

2.6.1.4. Introduction

The introduction should clearly state the problem under consideration and the current state of knowledge as it applies to the subject matter. It should also justify why this is an important area of study. It is in the introductory section that a review of the literature is usually presented. It is quite important that this review is as complete as possible. While the literature in most fields is substantial, and it is possible to miss important published articles, it can be embarrassing if one of these papers turns out to be of critical importance. In addition, it is important to realize that the relevant literature may not all be confined to English. At the conclusion of the literature review, a summary statement should follow which sets out the problem to be investigated and the purpose of the current research. This could be formulated as a question. It is here where the hypotheses might be presented as statements in the form of 'if *a* then *b*' (Section 2.5.1).

2.6.2. Materials and methods

These two sections deal with the materials used in the research as well as the methods used to arrive at the results.

2.6.2.1. Materials

This section is concerned with a description of materials used by the investigator; that is, the samples to be utilized. The materials section discusses the groups (samples) used, and how they were chosen. It details how each of the samples was collected and how the subjects were allocated to their respective groups. Along with a description of the samples, there should be a discussion of the experimental design used to test the hypotheses presented in the introduction. This refers to the number of and type of groups. For example, treatment groups and control samples, etc. Details such as age, sex, geographical locality, etc., are also specified. The total number of participants or specimens, assigned to each group needs to be listed. Descriptive statistics such as group means and standard deviations of some of the sample parameters such as specimen ages can be profitably presented here.

2.6.2.2. Methods

The second section is concerned with the techniques and procedures to be applied to those samples to produce results. The methods section, deals with the equipment utilized, the measurements taken and the reliability of these measurements. Variables, broadly defined in the hypotheses need to be carefully identified now and described. If applicable, dependent and independent variables should be distinguished. Novel or specially constructed equipment may need to be displayed with illustrations. It is essential that sufficient methodological detail be provided so that an independent investigator can replicate the results. In addition, there should be a discussion of intra- and inter-observer errors. Finally, it should be mentioned that the two sections, materials and methods are often combined into a single section called *Materials and Methods*; often there is considerable flexibility in how these two subsections are labeled.

2.6.3. Results and conclusions

These sections describe the results of the research and the conclusions, which are to be drawn from an analysis of the investigation. Finally, the end of the manuscript contains the references cited.

2.6.3.1. Results

It is here that the results are presented in various summary ways. The purpose of the results section is to provide enough information so that the conclusions drawn by the investigators can be understood. To make a convincing case sufficient data (generally in summary form) needs to be presented and arranged in an orderly and clear fashion. These may include graphs (in 2-D or 3-D) of the relevant variables. These may also be figures, which display the results, and/or tables, which show the results in tabular form. Although the order of presentation of graphs, etc., is at the discretion of the researcher, it is strongly suggested that the potential investigator consult the specific journal literature for examples of style. Since manuscript length is often limited by journals, this requires judicious choice with respect to the number of graphs, figures and tables that can be included in the manuscript. Descriptive statistics summarizing the data is usually placed first in the manuscript. This is followed by tables that display the results in terms of statistical significance. This is particularly important, as the goal of research is often the elucidation of differences between groups (*e.g.,* differences between sexes, treatment versus controls, etc.). Thus, the purpose of these statistical tests is to establish that the measured differences seen are not simply due to chance. Sophisticated statistical techniques, such as one of the various multivariate statistical approaches, may also have their place here. Note that in the results section there should be no attempt at relating the results into a broader context, that is the purpose of the discussion section.

2.6.3.2. Discussion

One purpose of this section is to relate the hypothesis presented in the introduction with the results that were subsequently found by the investigator. This interpretation represents an attempt to explain the results. The results may also be compared with the available literature. The purpose of comparing the work with that of other investigators is to place the current study into a wider context. Confirmation or denial of earlier results by others is often a relevant part of the discussion. Limitations of the study should also be discussed here as well as the applicability of the results to other populations. Carefully thought out observations of future research directions suggested by the current work are appropriate here.

2.6.3.3. Conclusions

This section, usually a rather brief one, may simply be a listing of the results found in the current investigation, often in summary form. At other times, it has been combined with the discussion section and serves as its summary.

2.6.3.4. References cited

The references cited (bibliography) section contains all the references used in the preparation of the manuscript. It is found at the end of the manuscript but before an appendix section (Section 2.6.3.5). Every reference that is cited in the body of the manuscript must be placed in the references cited section. In the biological sciences, the citation in the body of the manuscript is usually identified by either the last name of the first author followed by the year of publication or by a number identifying that citation. If an author format is used then the entry is enclosed by parentheses: *i.e.,* (author(s), year). The complete citation is then listed alphabetically in the references cited section. Examples of a particular citation style can be readily found in this volume. As indicated, two generally accepted basic styles are usually found in the references cited section. These are arranged either: [1] alphabetically by author or [2] by a numbered sequence starting with one. However, it must be cautioned that both: the type, author or number format, as well as the exact order of items in a citation tends to be journal specific. These format requirements will be described in the *Guide to authors*, which is readily available for all journals.

2.6.3.5. Appendices

The appendix section can be considered the absolute end of the manuscript. Included here is material that might be out of place in the body of the manuscript. Examples are lengthy mathematical derivations, complex and detailed data tables, the actual original data rather than just their means and standard deviations (generally required in master and doctoral theses), and other collateral data such as perhaps maps, computer programs, etc.

2.7. SOME FINAL COMMENTS

Most papers seen in the published literature are written in the third person. Prior to submittal, a careful check for spelling and syntax errors should always be conducted. The capability to carry out this check is generally incorporated into most word processors currently available for both desktop and portable computers. The entire manuscript should always be double-spaced to allow entry of reviewer, typesetter and editorial comments, unless a camera-ready copy is required. Figures, graphs, charts and tables should be laid out on separate sheets. Much of this material is in the form of black/white line diagrams. These diagrams were often manually created as India ink artwork in the past, often by commercial artists employed by educational institutions. Today they can be easily created on a PC with appropriate software that is readily available. Color is discouraged because of the expenses associated with publication. Legends for each chart, table or figure, etc., are listed separately from the diagrams. The order of the page layout of the manuscript that should be followed is as presented here starting with Section 2.6.1 and ending with Section 2.6.3.

Finally, it should be emphasized that quality research, if properly conducted (that is, ethically), is always based on the conscientious analysis of experimental data. Always approach research with a critical outlook, this requires rigor and skepticism. Do not, in your pursuit of the research goals, take shortcuts or accept facts, ideas, or theories without good sound reasons; subject your own thoughts to careful scrutiny. Avoid self-deception, bias, careless errors, etc. in all aspects of the research endeavor and you will be rewarded with the knowledge of having produced good work. Given an interesting hypothesis and using data, which has been collected using objective criteria such as [1] adequate samples, [2] gathered in accordance with randomization procedures to reduce bias, [3] analyzed with appropriate statistical procedures and [4] independently verified will, in all probability, be acceptable for publication.

KEY POINTS OF THE CHAPTER

This chapter has been primarily concerned with two issues that confront everyone at the beginning of their scientific career. Namely, how to initiate and conduct a research project and how to eventually publish the conclusions. The process of carrying out a scientific research project and eventually writing up the results has been outlined and the steps involved described in a logical fashion. The guidelines presented in this chapter should be sufficient for the beginning investigator in the biological sciences to successfully initiate, complete and publish the results of a scientific inquiry.

CHECK YOUR UNDERSTANDING

1. What are some of the major differences between the scientific disciplines and the metaphysical ones?

2. List the steps of the scientific method as presented in this chapter.

3. It is rationalism or logical thought that underlies all the sciences. Mathematics is primary and empiricism is only of secondary importance in understanding how science operates. Examine these two sentences and discuss their validity in philosophical terms.

4. What is the difference between testable and untestable hypotheses? Consider some examples of each and apply criteria that would allow one to differentiate between them.

5. What roles do variables play in the development of hypotheses?

6. What are the steps needed to write a paper for publication in a scientific journal?

7. What are the differences between experimental and observational studies? List the advantages and disadvantages of each.

8. What are the differences between building a qualitative hypothesis and constructing a quantitative hypothesis? Why does a quantitative hypothesis have advantages?

9. Utilizing an existing or proposed research project, write a 200-250 word abstract. In the case of proposed research, describe the results to be expected, and provide a justification of potential conclusions.

10. Can you define the following terms? Why are they important? *Null hypothesis, statistical significance, correlation, experimental group, control group, independent variable, dependent variable.*

11. What are the differences between a research protocol and the actual research procedures as outlined in the chapter?

12. In terms of explanation, what is the difference between a hypothesis and a theory?

REFERENCES CITED

Ayer, A. J. (1952) *Language, Truth and Logic.* New York: Dover Publications Inc.

Bailey, N. T. J. (1967) *The Mathematical Approach to Biology and Medicine.* New York: John Wiley.

Blandford, D. H., Warren, G. B. and Campbell, E. M. (1984a) Introduction to research planning. *J. Dent. Edu.* **48**:246-250.

Blandford, D. H., Warren, G. B. and Campbell, E. M. (1984b) Uses and content of a research protocol. *J. Dent. Edu.* **48**:298-301.

Brunette, D. M. (1996) *Critical Thinking.* Chicago: Quintessence Publishing Co.

Conant, J. B. (1951) *On Understanding Science: An Historical Approach.* New York: New American Library.

Corvi, R. (1997) *An Introduction to the thought of Karl Popper.* London: Routledge

Gould, S. J. (1996) *The Mismeasure of Man.* New York: W. W. Norton and Co.

Grant, E. (1996) *The Foundations of Modern Science in the Middle Ages.* Cambridge: Cambridge University Press.

Grinell, F. (1992) *The Scientific Attitude* (2nd Ed.) New York: The Guilford Press.

Healy, M. J. R. (1989) Measuring measuring errors. *Stat. Med.* **8**:893-906.

Herrstein, R. J. and Murray, C. (1994) *The Bell Curve.* New York: The Free Press.

Huff, D. (1954) *How to Lie with Statistics.* New York: W. W. Norton & Co. Inc.

Jensen, A. R. (1969) How much can we boost IQ and scholastic achievement? *Harvard Edu. Rev.* **39**:1-123.

Kuhn, T. S. (1957) *The Copernican Revolution.* New York: MJF Books.

Kuhn, T. S. (1970) *The Structure of Scientific Revolutions* (2nd Ed.) Chicago: University of Chicago Press.

Lestrel, P. E., Huggare, J. A. and Abbink, B. E. (1996) A longitudinal study of the cranial base in shunt-treated hydrocephalics: Fourier descriptors. *J. Dent. Res.* **75:**337.

McGuigan, E. J. (1978) *Experimental Psychology* (3rd Ed.) New Jersey: Prentice-Hall.

Manly, B. F. J. (1992) *The Design and Analysis of Research Studies.* Cambridge: University of Cambridge Press.

Marks, R. G. (1982) *Designing a Research Project.* New York: Van Nostrand Reinhold, Co. Inc.

Meyer, S. L. (1975) *Data Analysis for Scientists and Engineers.* New York: John Wiley and Sons.

Olson, R. (1982) *Science Deified and Science Defied.* Berkeley: University of California Press.

Parratt, L. G. (1961) *Probability and Experimental Errors in Science.* New York: John Wiley and Sons.

Pearson, K. (1911) *The Grammar of Science* (3rd Ed.) Re-issued by Meridian Books, Inc. (1957).

Popper, K. R. (1959) *The Logic of Scientific Discovery.* London: Hutchinson.

Resnik, D. B. (1998) *The Ethics of Science.* London: Routledge.

Runyon, R. P. and Haber, A. (1980) *Fundamentals of Behavioral Statistics.* Reading, Massachusetts: Addison-Wesley Pub.

Searles, H. L. (1956) *Logic and Scientific Methods* (2nd Ed.) New York: Ronald Press Co.

Singer, C. A (1959) *History of Scientific Ideas.* Oxford: Oxford University Press.

Thomas, J. R. and Nelson, J. K. (1996) *Research Methods in Physical Activity* (3rd Ed.) Champaign, Illinois: Human Kinetics.

Westmeyer, P. (1994) *A Guide for the Use in Planning, Conducting, and Reporting Research Reports.* (2nd Ed.) Springfield, Illinois: Charles C. Thomas.

Zolman, J. F. (1993) *Biostatistics. Experimental Design and Statistical Inference.* Oxford: Oxford University Press.

3. A HISTORY OF SCIENTIFIC MEASUREMENT

Men think it divine merely because they do not understand it. But if they called everything divine which they do not understand, why there would be no end to divine things.

Attributed to the Ionian pre-Socratics
(Quoted from Farrington, 1961:82)

Of the gods, I know nothing, whether they exist or do not exist: nor what they are like in form. Many things stand in the way of knowledge—the obscurity of the subject, the brevity of human life.

Protagoras (*ca.* 481-411 BC)
(In Diogenes Laertius Vitae Philosophicus IX, 51)

The supposition that the future resembles the past, is not founded on arguments of any kind, but is derived entirely from habit.

Treatise of Human Nature I, Part iii, Section iv
David Hume (1711-1776)

3.1. INTRODUCTION

The procedures outline in the last chapter, represent the collective experience of many thousands of researchers, often painfully gathered in an atmosphere hostile to science (consider the fates of Bruno, Galileo or Lavoisier) spanning more than twenty centuries. Consequently, it was felt that a brief digression into the historical background of science and the science of measurement would prove to be both instructive and interesting.

At the outset, it should be stated that a popular if naïve view of science is that it is simply the application of a procedure, called the scientific method; and, if it is successfully applied, everything in the modern world can be somehow explained. It is suggested here that this view represents an oversimplification. The purpose of this chapter is an attempt to correct this widespread and misleading impression. What science is cannot be readily separated from other endeavors such as the theory of knowledge (epistemology), or from social factors or even from aspects of theology. These three ideas are intimately bound up with time, that is, with history. The development of the scientific method is also related to the contributions by a number of extremely able individuals who not only affected the times in which they lived, but were also influenced by those times. It is for these and other reasons that this chapter surveys the idea of measurement within the broader context of the history of science, from its uncertain beginnings to the time of Sir Issac Newton (1642-1727) when science and the scientific method became firmly established.

It is, perhaps, instructive to begin this chapter with a broad query: Why is the study of science within a historical setting relevant? Two answers can be given. The first one is that a study of the early beginnings of science is of great importance for an understanding of modern science. The second related one is that the neglect of history and the place of science within it can have serious ramifications. Science and technology, increasingly influences the daily lives of citizens worldwide at a pace that continues to accelerate. On the one hand, socially effective use of scientific results has improved the well being of most citizens. On the other hand, the use of scientists during the Third Reich bears witness to political excesses. A specific example is the use of policies prescribed by National Socialism with respect to the application of eugenics. These were then used, in part, to justify the Holocaust. Unfortunately, what is not always realized is that scientists, in general, tend to have relatively little control over how their scientific discoveries are put to use by the leaders of society. Witness the development and use of the atomic bomb. The social ramifications of the use of science need to be widely debated in the public forum to prevent a recurrence of the excesses that seem to have characterized specific periods of the 20th century:

> In science, more than in any other human institution, it is necessary to search out
> the past in order to understand the present and to control the future (Bernal, 1971a:28).

While beyond the scope here, such issues dealing with 'morality' or ethics are of fundamental importance and may determine the very future of mankind. From a somewhat different perspective, namely with respect to the ethical practice of science, this aspect was briefly touched upon in chapter 2.

Although the emphasis of this volume is on morphometrics, the historical development and utilization of measurements cannot be viewed in isolation from other scientific endeavors. It is argued that the beginnings of morphometrics lie at the very center of developments that led to modern science and the scientific method. Consequently, this chapter is an attempt to connect the historical threads that led to measurement from before recorded history to the present. Moreover, it is argued that three threads are primarily responsible for the development of the foundations of science. These are: [1] the Greek experience, [2] the translation of Arabic texts in the 12th and 13th centuries, and [3] the development of western universities (Grant, 1996). Aspects of these three threads, which culminated in the scientific revolution of the 17th century, are very briefly explored in the following sections. Given the limitations of space, considerable selection and brevity was required. As a consequence, the material in this chapter is by no means inclusive, and while it is based on extensive reading of secondary sources, it nevertheless represents a personal view of the historical events that are presumed to have taken place.

3.1.1. Precursors of Science

In a sense, the precursors of what we call science and the scientific method can be already discerned in the Paleolithic (Old Stone Age), if not earlier. While this is perhaps less science and more technology, the beginnings are evident. The increased ability of early man to utilize eye-hand co-ordination coupled with rational thought led to the beginnings of science. This can be seen in the frequency of archeological assemblages of sophisticated tool kits consisting of arrowheads, spears, scrapers, knives, etc. To construct these, the toolmaker must have an *idea* of the implement in mind; this in turn requires a *design plan* and considerable practice or *experimentation*. Thus, through the making and using of tools, it was possible to learn about the mechanical characteristics of the products being utilized, which, in a rather incipient way, laid the groundwork for the physical sciences (Bernal, 1971a).

The development of tools is a visible expression of this, as well as the need to control and manipulate aspects of the natural world. This became possible because the natural world consisted, to a considerable extent, of regular phenomena that was predictable. It is from this conceptual base that the world began to be viewed as *mechanistic* as well as *rational*. Thus, early man learned to observe, interpret and act accordingly, for survival. It is from such behavior that the observational and descriptive sciences arose. This need to exert control probably proceeded in an order starting with the *physical* world (finding water, using fire). This was followed by the *biological* world (searching for game, locating eatable plants) and the *societal* world (locating ones place within the social group, and later in the universe).

Clearly, it is not surprising that it is in the societal world that we find the rise of magic, myths and eventually religion. Curiously, one could make a parallel case with the modern sciences. Knowledge of the older 'hard' or physical sciences arose first so they are the most developed. This was followed by the biological ones. This 'continuum' terminates with the newest and 'soft' or 'least reliable' sciences such as psychology, sociology, etc. Reasons for this are not hard to find. At least on one level, they lie in the increased complexity of the phenomena being studied. Some of these implications are taken up in Chapter 5.

3.1.2. Development of Language

It is, perhaps, not an exaggeration to suggest that the evolution of language was of primary importance for the increasingly deliberate control of habitat. It remains uncertain when, and how, language arose in the evolution of early man. Language facilitates the rapid transmission of information, often critical to survival. It also allows the handing down of culturally important information through the generations. It also facilitates the cultural transmission of rituals concerned with birth, initiation and death, rituals that with time develop into mythologies. It is important to emphasize that primitive culture consists of both the mystic and the rational as far as knowledge is concerned, and any separation would be quite difficult, if not impossible. These developing cultural traditions are a blend of both. However, more to the point here is that the development of language provides a conceptual framework, which is independent from the objects to which they refer (Bailey, 1967). Put in simple terms, language provides a vehicle for describing the external world.

However, reliance on language alone for description of the external world remains insufficient because of its impreciseness. That is, precise knowledge about the external world cannot be obtained or applied without at least the simple notions of counting and measurement. In other words, what this means is that a special language is needed for preciseness and this turns out to be mathematics. Mathematics tends to be unambiguous, understandable and capable of verification by any observer, subject to the caveat that one is suitably trained. This is why mathematics plays such a pivotal role in the sciences (Ziman, 1978).

3.2. EARLY BEGINNINGS OF MEASUREMENT

"Man is the measure of all things" attributed to Protagoras (*ca.* 481-411 BC) and credited to him by Plato in the *Theatetus* written the following century (Klein, 1988). Today this quote might be more relevant if it was modified as: "Man is the measurer of all things" given our technological age. Clearly, measurement is pervasive throughout modern society, whether in the hands of public or practitioners of the sciences.

The beginnings of the science of measurement lie in the distant past and what led to its initial development will in all probability, never be accurately determined. It presumably started with the gathering of knowledge. It is suggested here that measurement must have been an equally early development. Nevertheless, the threads that trace the idea of measurement, as practiced by early man (the rise of modern man, *Homo sapiens,* being in excess of a 100,000 years ago) are so faint as to probably remain undetectable. Thus, the beginnings of quantification are extremely difficult to discover. Suffice it to say, counting and measurement can be considered as primary activities and it is not unlikely that at some stage in the evolutionary development of early man, they may have been used, albeit in a very simple way. With the development of language, the concept of *number* begins to take on more meaning. Consider human body parts: we have one mouth, one nose, two ears, two hands, two arms, two legs, etc. The hand can be considered as unity, but it is composed of four fingers and a thumb. In general, the numbers one through four represent the limit of the human natural ability to count without aid. The eye is simply not a sufficiently precise numbering tool (Ifrah, 2000). Nevertheless, we do have the tools for counting, eventually leading to arithmetic (Sarton, 1952). In fact, these anatomical units of measurement make the earlier quote "Man is the measure of all things" quite understandable in a literal sense, even if quite removed from the intention of the original author. Moreover, from counting it is not far to measurement. However valuable these humble beginnings of arithmetic, counting was also essential in another area of human endeavor, numerology. The rise of numerology or magic numbers will be taken up later.

As Sarton cogently stated, the need for higher numbers than four or five led to the notion of *groups*. Thus, counting, if practiced at all, consisted of one, two, three, four and many. Five in particular became important as sets of five notches on a stick followed by a space could be readily recognized.[13] The use of vertical lines for one through four, seen in most of the early civilizations, has very ancient roots.[14] In fact, as Ifrah has noted:

> Perhaps the most obvious confirmation of the basic psychological rule of the "limit of four" can be found in the almost universal counting-device called (in England) the "five barred gate". ... used by innkeepers keeping a tally ... (Ifrah, 2000:7).

---

[13] One of the oldest surviving examples of counting by fives can be seen in a 7 inch-long stick that displays 55 deeply incised notches, of which the first 25 are arranged in groups of five. This specimen is of Paleolithic age (illustrated in Struik, 1987:11).

[14] Aramaic, Cretan, Egypt, Elam, Etruscan, Greece, Indus, Hittite, Lycian, Lydian, and Phoenician civilizations, at one time or another, all used vertical strokes for one through four, sometimes five (Ifrah, 2000).

This "five barred gate" system consists of the use of vertical strokes for the numbers one through four, which are then crossed with a horizontal line to identify the number five, this process being repeated for each set of five. Clearly, here was the beginning of a base (of five) for a number system. In fact, the use of the five digits of each hand (or feet), taken together, gives rise to the most frequently used number system, that is, the base of ten. While some American Indians used five and the Mayas used twenty, as the base of their number system, this need not detain us here. The use of such a base allowed the development of the four operations of arithmetic—addition, subtraction, multiplication and division. This can be made clearer with an example. Consider the number 19 (20-1). It is much more convenient to reduce the addition of two bases (of ten) by a digit, than exhaustively begin counting all the numbers up from one. The counting of multitudes of groups leads, quite naturally, from addition to multiplication. Multiplication is simply the addition of groups. Also more important, was that much larger numbers could be handled (Sarton, 1952, Conan, 1982). The simple numbers (1—10), and the symbols used to represent them, are surprisingly similar across cultures.

All ancient civilizations initially used parts of the human body for counting. When such measurements came into general use is unknown. Presumably, it goes back to early man. Thus, fingers, palms, feet and forearms became the first standardized measurements for lengths (Dilke, 1987). After all, even today one can still hear the familiar measurement of "a horse is so many hands high" or talk about the "rule of thumb". A particularly useful historical source for measurement in general and for numerous civilizations in particular, is the recently published work of Ifrah (2000).

Nevertheless, it should be noted that the processes outlined above did not progress very much for many millennia because they represented adequate solutions to the problems being encountered. It is only when conditions markedly change that the earlier solutions are no longer successful, marking, what has been called a paradigm shift.[15] The first beginnings of such change in worldview can possibly be identified as a revolution in food production with the development of incipient agriculture in the Near East.

It is this major change of life style, some ten thousand years ago in the *Fertile Crescent*[16] that led to the first real glimmerings of what we call science. It is at this time[17] that with the development of incipient agriculture, man was first able to settle down in stable communities. This was made possible, in part, because of the domestication of wild grains and the beginnings of husbandry, both of which are essential for the support of larger population aggregates. This in turn facilitated the specialization of activities and allowed for time that could be devoted to other than survival activities. For Neolithic agricultural techniques to prosper, fertile soil was required. This was found in lower Mesopotamia as well as in the rich riverine soils

---

[15] The use of 'paradigm' refers to a pronounced societal change or revolution in worldview. It is borrowed here from Kuhn (1970) whose usage was somewhat different.

[16] The Fertile Crescent is the geographical region in the Near East located roughly from the north of Israel to Iraq/Iran. It is composed of low-lying hills rich in soil, which allowed the development of incipient agriculture. The evidence for this incipient agriculture comes from archeological sites in this region. These include Jericho (~7000 BC), Jarmo (6500 BC) and a host of other sites. The initial beginnings of these sites are probably older, *i.e.,* closer to ~9000 BC (Mellaart, 1961:57-59).

[17] Technically, although not always consistent, this marks the shift from the Mesolithic to the Neolithic.

arising from the regular inundation of the Nile (McEvedy, 1967; Kinder and Hilgemann, 1972). It is with the development of the principal civilizations of Mesopotamia and Egypt, each with a distinctly different numerical system, that quantification as a procedure formally emerges. Additionally, it has been argued that at some stage, writing becomes an essential pre-requisite, if the development of science is to proceed. This is predicated on the need for permanent records to be stored as public archives (Ziman, 1978).

3.2.1. The First Civilization: Mesopotamia

Small bands migrated from the hilly flanks of the Fertile Crescent into the fertile river valleys of the Tigris and Euphrates by ~4000 BC. Population growth was sustained by an abundance of available food. As early as 4000 BC, the city of Eridu may have contained upwards of a thousand people (Mallowan, 1961). The oldest example of writing, a limestone tablet, was found at Kish and is dated ~3500 BC. This tablet also contained numerals (Mallowan, 1961; Olson, 1982). At this same time at Eridu, the period is identified as the Ubaid culture, there can be seen the first sacred structure or temple complex. It is in the next succeeding period of Uruk (~3500-3000 BC) that one can begin to trace people identified as Sumerians, who created the first full urban centers that are the forerunners of civilization. Present now in Sumer (~3200 BC) is the temple complex called a *Ziggurat*. In addition, by 3200 BC numerals appear on artifacts made during this period.

Mesopotamia's numerical system was re-constructed from the large number of Babylonian clay tablets that were found. It probably arose ~2900 BC and was fully developed by 2000 BC. The number system was based, but only in part, on the decimal system. Interestingly, it was also simultaneously based on the sexagesimal system (base 60).[18] Using the symbol ▼ for 1's and the symbol ◀ for 10's, we can represent the number 34 as ◀ ◀ ◀ ▼ ▼ ▼ ▼. After the number 59, a different system was utilized. To illustrate, our base ten system is defined by position so the number 345 contains a unit (5), a tens column (4) and a hundreds column (3), which is equal to 300 + 40 + 5, or $3x10^2 + 4x10^1 + 5x10^0$ in base notation. In a similar way the Sumerians would express 345 but with the base of 60, not 10; that is, 345 would be equal to $5x60^1 + 45x60^0$ or ▼ ▼ ▼ ▼ ▼ (space) ◀ ◀ ◀ ◀ ▼ ▼ ▼ ▼ ▼. A space determined column position here. Although, a place marker was initially not used to locate the unit column (that had to be inferred from the context), numbers smaller than unity (fractions) were utilized (Lindberg, 1992; Dilke, 1987). This system of counting is superior to that of the Egyptians (see below) in that it readily allowed for the use of fractions as well as incipient algebraic calculations. For example, given the product of two numbers and their sum or difference, find the two numbers.

Mesopotamian linear measures were based on the Sumerian cubit of 49.5 cm. This is based on a statue of Gudea, king of Lagash (2170 BC, now in the Louvre). From this and other evidence it has been deduced that 30 digits (each 1.65 cm) = 1 cubit. Other Sumerian measurements exist, for example, of area and of cubic capacity (listed in Dilke,

---

[18] We have retained the use of the sexagesimal system to measure time (60 minutes) and angles (360 degrees).

1987). However, it should be cautioned that such measures were not always standard but have differed depending on the region and time period.

3.2.2. Egyptian, Roman and Later Accomplishments

The rise of Egyptian culture is can be traced to ~5000 BC when Neolithic settlements begin to appear in this region. Early pre-dynastic developments can be dated to ~4000 BC. Subsequent developments in 3200 BC, lead to the first dynasty (Aldred, 1961). It is presumed that around this time, Egyptians developed a number system based on decimals. They employed a different character for each power of 10 (1, 10, 100, etc.). For example, using a | to represent 1's and a n to represent 10's, the number 34 can be written as ||||nnn. While earlier than the numerical system of the Sumerians, it is not as versatile. Although the system allowed addition and subtraction, division and multiplication become laborious, and generalized fractions, except those with the numerator of one, were unknown (Lindberg, 1992).

In Egypt, quantification can be seen in the development of astronomy resulting in the division of a year into 12 months each with 30 days to which were added 5 days. Hieroglyphs on tombs illustrate star positions from which the length of the night could be calculated. Egyptians also devised the 12-hour day and 12-hour night. Presumably, this information was of value to a farming community. In brief, Egyptian measurements of length were as follows: [1] 4 digits = 1 palm, approx. 7.5 cm; [2] 7 palms = 1 cubit approx. 52.3 cm (also a forearm length called a royal cubit) and [3] 100 cubits = 1 *khet* (rod), approx. 52.3 m. Other cubit measures seem to be non-Egyptian (Dilke, 1987).

One of the most important documents to be recovered was the Rhind mathematical papyrus written in Egyptian hieratic script dated from circa 1575 BC. It dealt with the area of triangles and the slopes of pyramids (Dilke, 1987). It was presumed copied by a scribe called Ahmos or Ahmose from an earlier papyrus dated 1849-1801 BC. While some understanding of right triangles was present with the use of the sides' 4:3 ratio, it did not lead to the use of the Pythagorean theorem. Also present was an approximation of π as $(16/9)^2 = 3.1604938$, which is too large by 0.019. Clearly, the Egyptians made use of basic geometrical principles in the construction of the pyramids, although the actual building procedures involved remain largely unknown.

Nevertheless, one can see in the earlier Egyptian developments the forerunners of the Roman and eventual English measurement system. The smallest measurement of distance was the finger's-breadth or Roman *digitus* derived from the Greeks. Four of these finger's-breadths became a *palm*, and four palms became the *pes romanus* or Roman foot (29.57 cm to 29.65 cm). The *pes* became the basis for distance, area and volumetric measures. It was in the late 2nd century BC that the Latin duodecimal system was adopted which divided units into 12 parts. Thus, according to Dilke (1987:26):

> The Latin word for one-twelfth, *uncia*, is derived from *unus*, so literally means unit. It has given the English language both 'ounce' and 'inch' (Old English *ynce*).

Augustus (27 BC - 14 AD) implemented the standardization of all rural units of measurement with those of Rome. The modern mile (1.6093 km) comes from the Roman *mille* or thousand. However, the Roman mile was equal to 5,000 feet, not 5,280 feet, which defines the modern terrestrial mile. The reason for this 'discrepancy' is that in

English practice, the foot (30.48 cm) was larger than the Roman foot.[19] Such measures as the *pes* and the *mille* became crucial in terms of land management and the building of roads throughout the Roman Empire. A particularly clever device developed by the Romans was the odometer. The odometer was attached to a two-wheeled cart. The rotation of the cartwheels, via a cam, allowed pebbles to drop into a container at regular intervals, thereby facilitating the measurement of long distances.

The distance from outstretched thumb to the nose tip became the yard; it is attributed to Henry I of England (1100-1135). This measurement, the yard, was eventually codified into a distance of three feet (91.44 cm) credited to Edward I (1272-1307). Edward I is also presumed to have defined the modern acre[20] as an area of 40 by 4 rods, which equals 43,560 square feet (4,407 square meters). Finally, it should be noted that widespread differences in lengths, areas, weights, etc., were the norm in antiquity. Preciseness of measures arose only when accurate instruments became available. For other measures such as weights and capacity, Dilke (*op. cit.*) should be consulted.

3.2.3. Developments on the Indian Subcontinent

Developments in ancient India are less well understood, in part, because archeological investigations have only been undertaken in this century. There is good evidence that at the time of Sargon of Akkad in Mesopotamia (reliably dated as 2350 BC) considerable trade reached the Mesopotamian cities from the Indus Valley in modern Pakistan (Wheeler, 1961). While the chronology is somewhat unreliable, the timetable for the origins of the early civilizations of the Indo-Pakistan sub-continent are ~2500 BC for the Indus civilization and ~1000 BC for the Ganges civilization. The Indus civilization is best known from two sites; one at Mohenjo-daro in Sind and the other at Harrapa, 400 miles southeast in the Punjab. Present was a script that is still largely undeciphered as well as different from that found in Mesopotamia and Egypt. Both Indian cities display sophisticated city planning. By about 1500 BC, both cities were sacked by invaders, possibly Indo-European Aryans, not to again recover (Wheeler, 1961). However, decline had already set in earlier, possibly due to exhaustion of the land, according to Wheeler. A later tradition, the Ganges civilization, is present from 1000 BC to 500 BC in the Ganges-Jumna basin, whether that tradition represents a secondary Aryan invasion is unclear.

The first well-preserved Indian contributions dealing with the sciences can be traced to the *Siddhontos* (300-400 AD). These books deal mainly with astronomy showing epicycles. They suggest the presence of Greek influence. The best-known achievement of Hindu mathematics is the positioning, or place-value, of our decimal system. It is thought to have arisen earlier in China. The first occurrence seems to have been in 595 AD. Present later was the dot to express the zero position, which dates to the 9th century (Struik, 1987). Eventually, the Hindu place-value system finds its way to the Islamic world (Section 3.4.1.)

---

[19] Latin 'mille', which refers to a thousand paces, is derived from one pace equal to five Roman feet.
[20] Acre is derived from the Latin 'ager', which is the size of a field plowed in the course of one day.

3.2.4. Rise of Chinese Civilization

The beginning of Chinese cultural traditions can be traced to ~4000 BC with the earliest known Neolithic site being Pan Po Tsun in Shensi Province (Watson, 1961). In contrast to the West, cultural stability with little change over time seems to have been the norm in China. The discovery of the capital of the Shang dynasty, based on excavations in 1927-1936 and continued after the Second World War, has reinforced the Chinese historical tradition of the beginning rulers in 2852 BC. The last ruler of the Shang dynasty was eventually overthrown by the Chou in 1027 BC who then founded a new dynasty (Watson, 1961). The earliest form of writing, as seen on oracle bones used for divination (dated to ~1500 BC), is quite similar to the script used today.

3.3. GREEK AND ROMAN SCIENCE

Sometime before 3000 BC, people began to occupy the Aegean archipelago, with settlements in Crete, the Greek mainland and in Asia Minor (Hood, 1961). The beginnings of a highly developed civilization in this region seem to have started in Crete. By 2000 BC, it flowered and eventually included writing (called Linear B). Cretan civilization can be said to be comparable with developments in Egypt and Mesopotamia at the same time. The presence of sophisticated architecture both in Crete (*e.g.,* the palace at Knossos) and on the mainland (*e.g.,* the tombs at Mycenae) seems to require the use of measurement, but the specific evidence is largely lacking. However, it should be noted that a Linear B surviving tablet (~1400 BC) does display a rudimentary number system (Hood, 1961). By 1100 BC, a precipitous decline of the Minoan culture becomes evident, whether due to outside invaders, such as the Greek Dorians, or coupled with natural disasters is a matter of some dispute. Nevertheless, the stage is now set for the rise of classical Greece where unprecedented events take place. These developments were to have a decided impact on the future course of knowledge; ideas that continue to resonate today.

It is convenient to break up the classic period of ancient Greek history into four time periods. The initial formative period is termed the pre-Socratic period, which is, in many ways, particularly interesting and scientifically significant. This is a period from approximately 600 BC to just before 400 BC. The second phase is from 400 to 300 BC, the philosophic period of Plato and Aristotle but also including the Epicurean and Stoic philosophers. The third interval is the so-called Hellenistic period from 300 BC to 100 BC initiated by the conquests of Alexander the Great. This period, also of particular relevance to science, contains the works of Euclid and Archimedes, among others. It is also the period of the foundation of the library at Alexandria, which was to last for 900 years. Finally, the fourth division is the Greco-Roman period, from 100 BC to 600 AD. This last period reflects a decline in Greek science as other spiritual and non-rational concerns associated with the rise of Christianity become dominant (Clagett, 1994). Nevertheless, it is also the period when Greek science passes to the Arabs, and through them, eventually to the Latin West.

To the Greeks, mathematics was considered a crucial aspect of education, while in the later Roman period it was viewed in primarily technological terms. At the outset, it should be emphasized that the term *science* has considerably different connotations when viewed in ancient terms. The ancient definitions (*scientia* in Latin, *epistomo* in

Greek) were applicable to *any* organized belief system whether it had to do with nature or not. Thus, an investigation of nature or what later became *natural philosophy* is indistinguishable at this time from magic, metaphysics or theology (Lindberg, 1992). The clear separation between science and religion did not begin to occur until the middle of the *enlightenment* of the 18th century, although hints of it were already present in ancient Greece.

3.3.1. The Pre-Socratics

The beginnings of philosophical inquiry leading eventually to science, as we know it today, have their roots in the Greek experience starting around the 6th century BC. This early period in Greek history is particularly relevant, as it is the first time in western history that a scientific viewpoint of the external world appeared. The earlier societies such as Babylonia and Egypt provided a certain level of theoretical knowledge (definitely necessary for future developments), particularly in the areas of mathematics, astronomy and medicine but did not attempt to provide explanations in naturalistic terms. Explanation tended to be defined in metaphysical terms. These civilizations developed strong centralized control in the hands of priests and kings, which led to monumental architecture (the Ziggurats of Mesopotamia, pyramids of Egypt), which require systematic organization of resources but at the expense of individual expression.

In sharp contrast, the beginnings of the Ionian experience were based on a system where political power was largely in the hands of a mercantile class, which considered development as progressive and essential to prosperity (Farrington, 1961). Wisdom or knowledge was considered useful and found its earliest expression in the numerous Greek colonies of Asia Minor that arose in the sixth century. It is with this Ionian school that there emerges the beginning of what was later to become *natural philosophy*. While influenced by the Egyptian and Mesopotamian cultures, they attempted to provide a purely naturalistic explanation to the universe as a whole. As Farrington has indicated, these early philosophers where not 'observers of nature' in an academic sense, but rather practical men whose minds focused on how things worked. They did this in the light of common experience without the need of metaphysical explanations. A word of caution needs to be inserted here. It is to be understood that, notwithstanding the considerable literature available on the Ionian experience, little in the way of actual writings or fragments of the period exists. What is known has come down via secondary and tertiary sources. This applies in particular to the works attributed to Thales and Pythagoras (Boyer, 1968).

Generally, these beginnings are said to have started with the school of Miletus and its founder, Thales (*ca.* 624-546 BC). Thales can be seen as a revolt against the earlier held views of cosmology as practiced by the Egyptians and Babylonians where the basis of the worldview rested on metaphysical concepts. Thales discarded those notions and attempted to provide a new worldview based on naturalistic principles. He accepted as a first principle the earlier viewpoint (held by the Egyptians), that everything was once water. For Thales water was the primary substance: all comes from water and to water all returns. Rather than accept a mythical explanation for the appearance of dry land, he suggested that it was formed by such natural processes as the silting up of the delta by the Nile. Thales believed that the earth was flat and floated on water. Furthermore, the heavens were also composed of water (a reasonable observation since it rains). While Thales may have gotten some things wrong, the recognition that the natural world can be explained by observation rather than a blind adherence to metaphysics was a major step forward and remains an essential principle of science. While Thales is considered the originator of a number of geometric proofs, the authenticity of this is difficult to confirm, as no first-hand records exist (Boyer, 1968). He evidently picked up knowledge of geometry from his travels to Egypt and Babylon. He is credited with: [1] having demonstrated that a circle is bisected by its diameter, [2] that the angles at the base of an isosceles triangle are equal and [3] that two intersecting straight lines produce opposite and equal angles. In astronomy he was the first to determine the sun's course from winter to summer solstice, and he is supposed to have measured the heights of the pyramids. Finally, the famous Socratic dictum of "know thyself" is attributed to him (Boyer, 1968).

Anaximander (*ca.* 610-546 BC), another resident of Miletus, objected to Thales' view that the earth was supported by water, since this raises the question of what supports the water. He attempted to provide a much more ambitious and detailed model of how the universe works. He proposed four elements. *Earth*, heaviest in the center, was covered by *water* over which there was a *mist* (clouds?) and surrounding all, *fire*. If we look upwards, we see holes (sun, stars?) through which fire can be seen. Especially intriguing, is Anaximander's view of biological evolution. He thought that fish preceded land animals and that man, accordingly, had once been a fish (Farrington, 1961).

The last Ionian philosopher from Miletus to be considered here was Anaximenes (*ca.* 550-475 BC), who suggested that only mist was the principal substance. While this seems a step backward compared to Anaximander, it, nevertheless, has considerable merit when considered in more detail. The idea is accompanied by the notions of rarefaction and condensation. That is, rarefied mist is fire and condensed mist becomes first water and then earth. He coupled heat with rarefaction and cold with condensation. Both ideas evidently were drawn from observation. The contribution of Anaximenes is that it provided an explanation of how a fundamental substance might exist in different states. However, none of the Milesian philosophers attempted to directly deal with the processes that led to change. This remained the task of Heraclitus of Ephesus (*ca.* 550-475 BC). For him everything was change: all things flow; nothing endures. For Heraclitus the principal substance was fire. The choice of fire probably represents an active agent of change for Heraclitus. He argued that the essential nature of things was always changing:

> All things come out of one, and the one out of all things. Change that is the only thing in the world, which is unchanging (Mackay, 1991:116).

Another thought also attributed to Heraclitus is that: "You cannot step into the same river twice".

3.3.2. From Pythagoras to Democritus

Another philosopher of particular importance from the Ionian school was Pythagoras (571-496 BC). With Pythagoras, the Greek philosophical tradition leads away from the naturalist or materialist view of the Ionians discussed so far, and returns to the metaphysical. Although born in Samos not far from Miletus, he eventually settled in Magna Graecia, what is now southern Italy. Pythagoras, like Thales, was said to have traveled to Babylon and Egypt. Because of the loss of documents, Pythagoras is also a very shadowy figure. The two words: *philosophy* (love of wisdom) and *mathematics* (that which is learned) are attributed to him. It is, nevertheless apparent that Pythagoras played a crucial role in the development of mathematics. He is presumed to have started a religious 'school' that resembled other cults present at that time but one that particularly emphasized mathematics and philosophy. While the motto of the Pythagorean School was "All is number", this is probably Babylonian in influence (Boyer, 1968). Moreover, the actual theorem of Pythagoras also has Babylonian affinities and the suggestion that the Pythagoreans first demonstrated it, cannot be verified.

The Pythagoreans viewed numbers as geometrical dots. They would inscribe figures into the sand. For example, using three dots one can create a triangle, with four a square. From this arose the triangular (1, 3, 6, 10, 15, ... ,) and the square (1, 4, 9, 16, 15, 25, ... ,) numbers. This led to the study of prime numbers, progressions and proportions (see Kline, 1972:30-34 for specific examples). Besides the apparent mathematical sophistication, this number system also led in another direction, to the use of numerology. While not original to the Pythagorean School, this mysticism can be seen in the development of the sacred numbers. Four was a special favorite as well as ten. Ten was a triangular number composed of three sides of four dots (Kline, 1972). In later times, ten was viewed as a perfect number, the symbol of health and harmony. The number seven was held in special reverence, possibly because of the seven wandering 'stars' or planets (that today account for our days of the week). Each number was associated with particular properties. Four, mentioned above, was associated with justice or retribution; hence, the adage: "squaring of accounts". Five is the number of marriage or union, since two is female and three is male, and so on. In contrast to the Egyptians with their unit fractions and the Babylonians with rational fractions, for the Greeks number was limited to integers, although relations between integers, ratios, were acknowledged (Boyer, 1968). However, the presence of *irrationals* (their term) generated a crisis since; for example, the $\sqrt{-1}$ cannot be expressed by an integer. The Pythagoreans eventually applied these mathematical concepts to astronomy. For the Pythagoreans, matter became numbers. Using observation but mostly *a-priori* [21] reasoning, they produced a more sophisticated cosmological view of the universe compared to the earlier Ionians. In the center was a mass of fire around which revolved the earth, sun, moon, the fixed stars and five planets. This cosmological view was not altered until Ptolemy (Farrington, 1961). Finally, the contribution of Alcmaeon of Croton (*ca.* 500 BC) should be mentioned because it shows that, like the earlier Ionians, members of the Pythagorean School used observation as well as *a-priori* ideas to arrive at knowledge. He is credited with having conducted animal dissections and discovering the optic nerve. Moreover, he recognized that the brain was the center of sensation. He is

---

[21] The term *a-priori* refers to 'before experience' or outside of experience, *i.e.,* not empirical.
Mathematics is an example of *a-priori* knowledge since it is independent of empirical observations.

considered to have laid down the foundations of experimental physiology. This issue of sense-experience will become a major philosophical theme with time (Farrington, 1961).

The major contribution of the Pythagoreans, apart of the mysticism of numerology, was that mathematical principles, for the first time, became closely tied to philosophical ones, rather than to purely practical ones, as had been the earlier tradition. In one sense, this connection has been continued to the present day. The Pythagoreans not only represented a break with the earlier Ionian naturalistic tradition, but they became one of the major philosophical influences on the thought of Plato and Aristotle (see section 3.3.3).

Nevertheless, the Pythagoreans had their detractors. One of these was Parmenides of Elea (fl. 501-492 BC) who was the most prominent member of the so-called Eleatic School. His views were diametrically opposed to the concept of change advocated by Heraclitus. He proposed the radical position that change—all change—was on logical grounds, impossible: that out of nothing comes nothing (Lindberg, 1992). Furthermore, he questioned the idea of opposites, such as being and not being (the void), which led him to the logical principle of contradiction. A principle directed against the notion of a fundamental change in matter as advocated by Anaximenes. Thus, the notion of water changing to earth was a contradiction since matter remains unchanged. Nevertheless, his focus on the distinction between what can be experienced through the senses and through reason is fundamental. It has influenced philosophy down to modern times. He did not claim that reality *was* thought; only that it could only be apprehended with thought (Collinson, 1987).

Parmenides pupil Zeno (fl. 464? BC) attempted to further extend this doctrine of reality with a set of proofs. The central issue here was whether ultimate reality was even knowable. The question is not whether observation per-se, *i.e.,* sense data, can be obtained, but rather whether such sense-experience really represents reality and can be trusted. This notion became the basis of *idealism* (as opposed to *realism*). It will play a major part in the subsequent history of western philosophy. Parmenides and Zeno, argued, on rational grounds, that what was experienced was an illusion, not reality. Reality could only be approached with the use of rules of logic. These ideas will be again encountered in the philosophy of Plato.

Counter arguments were produced by Empedocles of Agrigentum, Sicily (*ca.* 500-430 BC) and Anaxagoras of Clazomenae, Asia Minor (*ca.* 488-428 BC). They both argued that while the senses may not be perfect, they are useful guides that can offer a glimpse of reality, however imperfect. It is from Empedocles that we get the model of the four primary elements: Earth, Fire, Water and Air (which replaces Mist). These four elements will become the accepted cosmological framework until the advent of the Renaissance. His replacement of mist with air was based on observations and experimentation, making sense-experience essential in the search for knowledge. Anaxagoras, of the Ionian school, also supported the idea of observation but recognized the limits of sense perception (Farrington, 1961). He proposed the notion of first principles based on 'seeds' that contain a little of all the qualities, which we can recognize with sense-experience. Thus, in both Empedocles and Anaxagoras, one can see the first steps taken towards the development of an atomic theory.

Leucippus (dates unknown) and Democritus of Abdera (*ca.* 460-370 BC) whose fragments survive represent the pinnacle as well as the end for the time being, of the naturalistic inquiry about the nature of the universe, first initiated by Thales. The atomic theory proposed by Leucippus and Democritus can be seen as a response to the problems posed by the Eleatic School. The Eleateans argued that ultimate reality was one, whole, motionless, uncreated, and that *not being* was impossible. A difficulty with this view is that such a model of ultimate reality is difficult to describe when faced, experientially, with change and motion. Leucippus and Democritus avoided this difficulty by proposing a different view. They proposed a return to the earlier dualism. That is, *not being* (the void) was as real as being. In addition, that 'non being' had to exist because it was necessary for motion (Collinson, 1987). Thus, the void (or cosmos) was composed of an infinite number of indivisible, unchanging particles that, in the aggregate, formed larger bodies that become the visible entities seen in the sensible world. This also hints at the idea of the infinity of space. The Pythagoreans viewed number as points in the sand, while these points have mathematical content, they cannot be used to build a universe. Democritus provided the units on which a universe can be constructed. Thus, the atomic theory was a brilliant hypothesis. The impact of this view of Greek philosophy on western society has been considerable. The discoveries of John Dalton in the early 19th century led to a re-evaluation of the atomic theory as proposed by these pre-Socratic Greeks (Farrington, 1961).

Besides realism and idealism noted earlier, two other alternative views can now be discerned. Waddington (1959) has distinguished these two views: one dealing with *things* and the other with *process*. The first view is attributed to Democritus who devised the word 'atom' for these invisible and unchanging particles of matter. Clearly, in view of quantum theory, this is not exactly the modern view of atomic particles, but it is, in one sense, not far from the modern view. The second view, change or process, was championed by Heraclitus already discussed. Science by this time had achieved remarkable advances, of which two major accomplishments can be singled out. These are: [1] explanations of natural phenomena without the need for supernatural intervention, and [2] the beginnings of observation and experimentation. Themes that will resonate sometimes very faintly but never be totally lost, over the span of the next twenty-five centuries.

It is particularly regrettable that of all the pre-Socratic achievements, the majority of the writings of Democritus were, unfortunately, lost. By the middle of the third century BC, the contributions of these philosopher-scientists were largely in decline, as other philosophical movements became increasingly predominant.

3.3.3. Platonic and Aristotelian Philosophy

With Plato (427-347 BC), philosophic principles become increasing divorced from experimentation. Apart of the Hippocratic writings, which are difficult to identify with specific authors; it is with Plato that we have the first extensive philosophical discourses that have survived. Plato, a student of Socrates who, it is presumed, did not record anything, wrote the famous dialogues. After Socrates' death, Plato encountered the Pythagorean philosophers, he then returned to Athens to form his own school or Academy in 388 BC (Lindberg, 1992). A very brief examination of Plato's philosophy reflects a rejection of the realism of the pre-Socratic view in favor of Idealism. For other aspects such as the beginnings of politics, ethics, etc., the interested reader is referred elsewhere for the extensive literature that is available. Here the focus is confined to Plato's theory of knowledge and how it impinges on science. With Plato, the duality of whether knowledge is obtained via the senses or by reason was re-examined. Plato does not dismiss the senses altogether as Parmenides did:

> The simple sensations which reach the soul through the body are given to animals and men by nature, but their reflections ... are slowly gained ... by education and long experience (quoted in Farrington, 1961:108)

Nevertheless, Plato opted on the side of reason, for him ultimate reality was unchanging and consisted of *Forms* or ideas. The real world, as it is perceived by the senses, is a rather poor copy, or shadow of these ideal, perfect Forms. For example, there exists a perfect circle. The circles that one draws, or perceives as sense data, are simply imperfect copies. He does not ignore change (seen in the real world of experience), but he relegates it to a 'lesser' reality. He felt that they played a role in leading one toward ultimate reality (Lindberg, 1992). His Forms or ideas can be viewed as the development of classes, such as white dogs and black dogs (sense-data), based on shared properties, belong to the class of dogs (the perfect dog). In brief, Plato's approach is one of rationalism as opposed to empiricism. Plato took over the four elements of Empedocles, earth, water, fire and air, but under the influence of the Pythagoreans, attempted to quasi-mathematize nature. That is, he reduced the elements to geometrical figures (the Platonic figures). In Plato we first encounter in, formal terms, the idea of form. It will be subsequently shown that this concept of form will go through considerable modifications and will play a central role in morphometrics.

Plato's astronomical view, with natural principles now controlled by divine principles, also reflects the Platonic Ideal. That is, the stationary earth is surrounded by the sun, moon and planets (re-introduced as celestial gods), all of which move in perfect uniform circular paths (Farrington, 1961). In all this, Plato represents a reactionary view to the earlier Ionian philosophical tradition. The emphasis on the Platonic Ideal shifts the emphasis back onto the individual soul and leads to the restoration of the gods. This set into motion an epistemology that led to metaphysics, to Christian doctrine and continues to have considerable influence.

The work of Aristotle (384-322 BC) has also has survived; it is extensive and collectively would fill a bookshelf. The longevity of the Aristotelian tradition, for over a millennium, arises because the construction of medieval Christian theology was based, in part, on Aristotle's works. In 335 BC, he founded the Lyceum as an alternative to Plato's Academy. The Lyceum was eventually associated with the Peripatetic school of thought (from *peripatos*, a colonnaded walkway). Other schools also flourished, such as the Stoics and the Epicureans (see below).

Aristotle's contributions were substantial. We focus here on just a small part of his epistemology, that part which was directed toward science. While influenced by Plato, he eventually rejected the Platonic Ideal, *i.e.,* that forms had separate existence, as not being in accord with experience. While he accepted the idea of classes and greatly extended it, the properties that characterized dogs resided for him, within the dogs and did not have a separate existence. However, this raised epistemological questions about what ultimate reality was. Consider a warm rock. It (the rock or subject) has a property (in this case warmth). All subjects have properties and a subject cannot exist without properties. Biological organisms, *e.g.,* consist of a certain shape and structure. That is, bone, skin, tissue, organs, etc., all organized along certain principles. Thus, these two aspects: shape (or form) and structure (or matter) are central to Aristotelian thought[22] (Barnes, 1982). Aristotle's answer as presented in his *Physics*, was to consider that reality consisted of two things: *form* and *matter*. These ancient concepts of form and matter continue to occupy a place in philosophy well into the Renaissance and still resonate today. Aristotle maintained that all knowledge was empirical; that is, gained from perception by the senses. Thus, Aristotle's epistemology is directed toward viewing nature and change; that is, the world, as sense data, if matter changes then form changes. This was now in direct contrast to Plato's view of knowledge composed of external forms, only knowable by reason or philosophical reflection. Moreover, while both Plato and Aristotle were concerned with universals, for Aristotle one started with individual observations. Thus, for Aristotle, knowledge begins with sense data or repeated experience, followed by memory, which leads to 'intuition' from which one can then discern the universals (Lindberg, 1992). Aristotle viewed all phenomena, whether physical or biological, subject to certain fundamental laws involving *matter* and *motion*. For Aristotle motion affects matter by determining its transition from *potentiality* to *actuality*. An example might be: seeds contain potentiality leading to actuality with the growth of plants. It is important to realize that the concept of motion is today associated with change in location (locomotion). Aristotle had a much broader conception in mind. If motion implies change then the potential to change can lead to actual change, hence the concepts of potentiality and actuality. In Aristotelean terms: "Change is the actuality of the potential *qua* such". A sentence often cited for the Aristotelean definition of motion (Barnes, 1982). Space limits a detailed discussion of this and other interesting aspects of Aristotle's work. The reader is encouraged to pursue the extensive literature available.

The Aristotelian world was an orderly and organized world in which everything develops toward an end (*i.e.,* it was *teleological*). He was interested in causation, like so many of the ancient philosophers, and proposed the notion of a First Cause or Prime Mover, which eventually evolved into the Christian notion of God. However, in general, he avoided first cause arguments, arguing that the universe was eternal and had no beginning, a point of some embarrassment for later scholasticism. His view was not the world of the Eleatic atomists, mechanistic and random. His cosmology, was similar to Plato's, he accepted the four elements of Empedocles—earth, water, fire and air, but added a fifth one to the heavens, ether. He viewed the earth as a sphere, which was later largely accepted by medievalists, the idea of a flat earth was a much more recent invention[23] (Lindberg, 1992).

With respect to the motions of the heavenly bodies, Eudoxus of Cnidnus (*ca.* 400-347 BC), a student of Plato, had initially suggested twenty-seven spheres to account for such movements as retrograde motion. He was followed by Callippus of Cyzicus (*ca.* 370-300 BC) who proposed thirty-four spheres. Aristotle modified the circular motion

---

[22] Note that for Aristotle as well as others since, form was interchangeable with shape. See Chapter 6 for a re -evaluation of the concepts of form and shape.

[23] The revival of learning in the 12th century Europe coupled with the availability of Ptolemy's astronomy and Aristotle's cosmology, resulted in a wide acceptance of the fact that the earth was a sphere, in spite of popular accounts to the contrary (Hall, 1994).

theory initially proposed by them, and increased the number concentric spheres that account for planetary motion to fifty-six. As Aristotle's cosmology was eternal, unmovable, planetary movement had to be accounted for. To handle this problem, Aristotle again proposed a *Prime mover* a sort of living deity expressing the highest good, not unlike the traditional Greek Gods. For Aristotle each planet had its own prime mover.

In matters of biology, Aristotle's contributions were also extensive, over four hundred pages in modern translation (Lindberg, 1992). As a justification for the study of zoology, he indicated that it contributed to the knowledge of the human form. This was predicated on the close resemblance between human and animal natures. He saw purpose in the animal kingdom, which refuted the notion of nature being the work of chance alone. That the Aristotelian concept of science has a modern ring can be readily seen with his view that biology was both descriptive and explanatory. He considered the explanation of biological phenomena as the ultimate goal that depended upon the gathering of biological data (Lindberg, 1992). He invented comparative anatomy by using the parts of the human body, both external (head, neck, arms, legs, etc.) and internal (brain, digestive system, sexual organs, lungs, heart, etc.) as a standard with which other animals could be compared. Perhaps his greatest accomplishment was in descriptive zoology. He classified more than 500 different species of animals. Both structure and behavior are described in considerable detail, often based on dissection. He devoted considerable attention to the problems of classification. He displayed meticulous observation such as the incubation of bird eggs. In his worldview, each organism is constituted of matter and form; that is, matter consisting of the various organs that make up the body, and form being the organized principle that molds these organs in a unified organic whole (Lindberg, 1992). Finally, with respect to the concept of a soul, Aristotle identified it with life; that is, all life forms possessed souls. This is a much broader definition than the more narrowly prescribed one later advocated by the Church. There is much more to Aristotle's biological system, nevertheless given the space limitations here, the interested reader is again directed to the copious literature that is available.

In sum, the Aristotelian philosophical system represents an astonishing accomplishment. Aristotle provided answers to the major philosophical issues of his day. He re-interpreted the major problems posed by both the pre-Socratics and Plato. In addition to the biological contributions already mentioned, he opened new areas of inquiry with the analysis of natural phenomena. His personal observations single-handedly contributed new knowledge. He commented on a dazzlingly large array of physical events, consisting of comets, shooting stars, rain, thunder and lightning. These also included geological considerations such as earthquakes, the list is almost endless. He also developed a theory of light and vision.

Just how does morphometrics and quantification fit into this scheme? The emphasis on form (or matter) and motion, as outlined above, clearly reflects Aristotle's recognition of the primacy of sense data. Sense data continues to be the basis of modern morphometrics. Aristotle viewed mathematics as separate from the natural sciences or physics, for him mathematics was an abstraction of sense data (Lindberg, 1992). This view contrasts sharply with that of Plato who, borrowing from the Pythagoreans, viewed everything in geometrical, not material terms. These two divergent views of the relation of nature and mathematics have continued to occupy natural scientists to the present. Thus, we need to look at how such mathematical ideas were translated into practical terms. We will return this topic in a moment.

3.3.4. Greco-Roman Achievements

After Aristotle's death, the Lyceum, which he had founded, was taken over by his colleague Theophrastus (*ca.* 371-286 BC) who held the position for the next thirty-six years. Theophrastus shared Aristotle's outlook and commitments, largely resulting in a continuation of the tradition. He, however, questioned aspects of Aristotle's natural philosophy, especially his notion of teleology. He indicated that not all features of the universe serve a purpose and that there seemed to be a substantial random element in the world (Lindberg, 1992). Unfortunately, most of his works have not survived. The next head of the Lyceum was Strato (dates unknown) who held the position from 286 to 268 BC. He also continued and expanded the Aristotelian tradition, again none of his works survived intact. Connections were eventually established between the Lyceum and the new museum or library at Alexandria, Egypt where, after Alexander's death, the Ptolemaic dynasties were beginning. Evidence for this can be found in the fact that Strato spent some time at the Ptolemaic court. Given that Athens was in decline, the library at Alexandria eventually became the primary institution in the connection between Greek thought and the subsequent Roman and medieval periods. Nevertheless, as far as the Peripatetic School was concerned, while Aristotelian ideas continued to be debated, little in the way of significant advances occurred. The Lyceum stopped functioning in the early part of the 1st century BC when Athens fell to Sulla and the rolls or volumes were shipped to Rome.

Besides the Lyceum, two other schools need to be briefly mentioned, as these became serious rivals of Platonic and Aristotelian thought. These are the Stoics and the Epicureans, while generally opposed to each other, they, nevertheless, shared a few philosophical elements. The aim of philosophy, according to Epicurus (341-270 BC) was to attain happiness. Achieving happiness is not to be construed in modern terms such as materialism and hedonism, but rather as a philosophy, which eliminated the fear of the unknown and the supernatural (Lindberg, 1992). Epicurus's natural philosophy was borrowed, in part from the atomists, and led to the world being mechanistic, in that everything is the result of mechanical causation. There is no life after death, no divine being, no final causes, etc. This view is in direct contrast to the Aristotelian concept of teleology. However, Epicurus was also opposed to sense data arguing that it was untrustworthy. Zeno (*ca.* 333-262 BC), not to be confused with the Zeno the student of Parmenides, was the founder of Stoicism. Stoics believed that happiness could only be attained through harmony with nature; thus, living in harmony, required knowledge of natural philosophy. Both philosophical schools were materialistic, denying the existence of non-materialist philosophy such as that of Plato. Both movements were to influence philosophical developments in the 17th century.

We now turn to other concerns, raised earlier (section 3.3.3), having to do with the application of mathematics to nature, starting with the Hellenistic and ending with the Greco-Roman period. As mentioned earlier (section 3.3.2), in contrast to the Egyptians and Babylonians, the presence of irrational numbers presented a major obstacle. This may have persuaded some Greek mathematicians that numbers were unsuitable for the representation of reality, leading instead to the formal development of geometric principles with proofs. This sets the stage for the greatest geometrician that the Greek experience ever produced. Euclid (*ca.* 300 BC) not only codified earlier developments in his thirteen volumes of the *Elements,* but for the first time developed a sophisticated axiomatic system that is still in use today. He began with a set of definitions: a point ("that which has no part"), a line ("length without breadth"), a surface, right, acute, and obtuse angles, parallel lines, etc. These definitions form the basis of Euclidean propositions and their proofs. These do not need to detain us here, as they are familiar to everyone with a modern secondary school education. The method became the standard *modus operandi* for science until the end of the 17th century (Lindberg, 1992). The influence of Euclid is still being felt today. It continues to unconsciously shape the thoughts of not only laymen but also the practitioners of modern science. Nevertheless, it needs to be noted that proof of Euclid's 5th postulate; namely, the parallelism of lines, was already being questioned in the 17th century (Bonola, 1955), leading the way toward non-Euclidean geometry. Non-Euclidean geometry not only allowed simplification of Einstein's theory of relativity, but also comes into play in the deformation methods of morphometrics (Chapter 8).

Euclid was followed by other scholars of equal aptitude. These include Archimedes (287-212 BC) and Apollonius of Perga (*ca.* 262-190 BC). Archimedes, apart of his famous water displacement experiment that led to having said: "I have found it" (*Eureka*), also made major contributions to both to physics and mathematics, as well as engineering. He developed methods by which the area of a bounded parabola and the surface area and volume of a sphere could be computed. Archimedes also calculated a closer estimate for π, between 3.1409 and 3.1429. He also seemed to have been aware of the scientific method and used it in his experimental work (Farrington, 1961). Apollonius is credited with a book on conic sections in which he studied the ellipse, parabola and hyperbola and provided a new approach in terms of their generation.

However, Greek examples of quantification were largely limited to contributions in astronomy. Aristarchus of Samos (*ca.* 310-230 BC) in an extant work, *On the Sizes and Distances of the Sun and Moon*, clearly applied quantification and measurement, although the figures were shaky and in need of correction (Farrington, 1961). The quantitative work of Eratosthenes (284-192 BC) supplied the lack of standard units in the work of Aristarchus and provided the first estimate for the earth's circumference at 250,000 stades, or 25,000 miles, a surprisingly accurate estimate (Farrington, 1961). Klein (1959) puts this value at 24,000, which is still, nevertheless, a close approximation.

With respect to cosmology, the acceptable model of the universe had been proposed by Eudoxus and accepted by Plato. Since the 4th century BC, it had been recognized that the sun the traverses the ecliptic once a year and the moon orbits once a month, both east-to-west. In addition, it was known that the planetary motions of the 'inner planets' never strayed very far from the sun. More troublesome was so-called *retrograde* motion since the earth was in the center of the universe. This was the so-called *homocentric* or *geocentric* concept of the universe. In fact, Farrington suggests that other factors were available that raised questions with the homocentric theory, questions that evidently, Eudoxus conveniently sidestepped. Such as the varying brightness of planets and the nature of eclipses, some total some annular, which hinted at varying, not circular orbits. Moreover, Aristarchus had proposed a heliocentric system with the sun in the center and the earth as one of the planets, although he was anticipated in this by Heraclides of Pontus (388-310 BC). Copernicus (Section 3.5) was well aware that he was unearthing a theory initially proposed by Aristarchus. Nevertheless, the heliocentric view was too radical since it shook the accepted convention about the earth being the center of the universe. In addition, it must not be forgotten that, for most Greeks, the heavens were divine and home to the gods. Thus, societal considerations also strongly mitigated against the replacement of the accepted homocentric theory by a heliocentric view.

Astronomical considerations nevertheless, were actively pursued with contributions by Apollonius and Hipparchus (dates unknown). These records have disappeared and we are dependent on what Claudius Ptolemy (*ca.* 100-170 AD) reported. Ptolemy (not a member of the ruling dynasty, but a citizen of Alexandria) was associated with the Library of Alexandria. He is credited, along with Hipparchus, for the development of spherical trigonometry, initially applied to astronomical questions. His compendium, *Mathematike Syntaxis* (Mathematical Compilation) later known as *He Megiste Syntaxis* (The Greatest Compilation), is also known by the corrupted Arabic as the *Almagest*. In his search for explaining the regularity of planetary motions, he introduced a number of models. The first one is the *eccentric* model, which is sufficient for dealing with simple cases of motion, such as the sun. However, to deal with retrograde motion, Ptolemy then had to introduce the *epicycle-on-deferent* model. An outer planet would move along the deferent (larger orbit) until it reached *retrograde motion* (slowing down, moving backward) at which point it would travel along the epicycle (small orbit) until it began to move forward again. During this process, the epicycle is also moving along the deferent. This rather elaborate system was devised to predict planetary motions. Ptolemy also developed an additional model to account for other planetary motions, but this need not detain us here. These models are dependent on the notion of uniform circular motion, being the simplest motion, a view also held by Hipparchus.

These models are firmly based on sense data, *but grounded in a particular worldview*. In addition to the reasons alluded to above, the fact that planetary positions could now be predicted with some accuracy also explains the longevity of the Ptolemaic theory. The Ptolemaic view was to last for over thirteen centuries, until the advent of Copernicus system. What is especially relevant to realize about Ptolemy, is not so much the legitimacy of the proposed theory, but rather the approach. He set up a program that has a definite modern ring to it. It is based on descriptive observations to which is coupled an abstract analytic framework, or mathematical model. All with the purpose of trying to discover the nature of the universe, what today we would call explanation. While he was, undoubtedly, not the only Greek to begin to practice science in this fashion, he can be considered as a singularly good example.

In summary, it should not be forgotten that the period starting with Plato and continuing beyond Aristotle to the Stoics and Epicureans was marked by profound social changes. The old mercantile class with a practical bent that had allowed for initial development of science in 6th century BC Ionia had been replaced. These historical developments can be traced from the 4th century BC to the 1st century BC. They include the rise and fall of the Age of Pericles, the rise of slavery, and expansion culminating in the Hellenistic Empire of Alexander the Great, and the eventual eclipse of Greece leading to Roman domination. Times became increasingly insecure and unstable leading to, at various times, political control with obsessive laws. In a slave society, scientific ideas were restricted to a small minority of citizens. Thus, citizens used words (reasoned) while slaves carried out deeds (work). Plato, in fact, had little use for what we would call applied science. Concomitant with the exclusion of the majority of the population, science became separated from applied or practical undertakings. This separation, in part, led to its decline, a decline from which it did not recover until the Renaissance (Farrington, 1961).

3.4. THE HELLENISTIC PERIOD INTO MEDIEVALISM

While Roman expansion, especially in the old Greek provinces, continued to reflect Hellenistic influences, it initiated a period of scholarship that respected Greek knowledge, but offered little that was new or unique. The focus was on synthesizing the Greek intellectual achievement and making it available for a new educated Latin audience. A good example of this is a work by Pliny the Elder (23-79 AD) entitled *Natural History,* which is a compendium of 20,000 'facts' drawn from over a 100 authors (Lindberg, 1992). Science or natural history as the Romans understood it, was now of a limited, popularized version. What was accepted was logic and rhetoric that had use in the law and politics, while epistemological and metaphysical considerations, anatomy, etc., were of little interest. With few exceptions, Roman achievements lay elsewhere, in political administration and engineering.

Besides Aristotle, another individual needs to be singled out for contributions to biology and medicine, and indirectly to morphometrics. This was the work of Galen (129-216 AD). His work was to influence medicine for over 1,400 years in both Byzantium and the Latin West. Of the 300 or so manuscripts attributed to him, approximately half have survived. By 500 AD, his works had become accepted knowledge, by the 9th century; they had been translated into Arabic. By the 11th and 12th centuries, Galen's works had reached the Latin West. They eventually became the core of medical learning at the newly developed medieval universities.

Galen developed his anatomical knowledge with the dissection of monkeys. Based on this first-hand experience he advocated dissection to improve surgical skills as well as for research. Although seriously hampered by the dictum against human dissection, which led him to make errors about human anatomy, he regarded the study of anatomy as the foundation of medical knowledge. In physiology, Galen held the earlier Greek view that human health depended on a balance between the four bodily fluids (humours), which were blood, yellow bile, black bile and phlegm. According to Galen, blood was formed in the liver, carried by the veins to the rest of the body and then somehow transformed there into various tissues. Galen's system was eventually overthrown in the Renaissance (Chapter 4), with the appearance of observations that were more accurate and the initiation of experimentation. So completely has Galen's influence dissipated, that it was only in 1952 that the first complete English translation of his works was published (Bernal, 1971a).

By the end of the 2nd century AD, the generally stable conditions that had favored scholarship had declined with the increase in disturbances, civil wars, and economic disasters. The end of this period of comparative stability can be identified with the death of Marcus Aurelius in 180 AD. By the 4th century, AD conditions had deteriorated to such an extent that the Roman Empire was barely recognizable as single political entity. Communication had practically ceased with the division of the Greek East and the Latin West into separate administrative units. One of few scholars at the time, Boethius (480-524), attempted to render as much as was still available of the Greek sources into Latin.

The increasing role of Christianity must now be briefly mentioned. Christianity grew into a major religious force by the 3rd century AD and became the state religion by the 4th century. The common assumption has been that Christianity was anti-intellectual, preferring faith instead of knowledge, and this was responsible for the decline of science from which it did not recover for a millennium. This is an oversimplification as the picture is considerably more complicated (Lindberg, 1992). While the Church was on a proselytizing mission, it also recognized that if the Bible was to be read, literacy had to be encouraged; eventually though learning was cultivated only if it led to spiritual pursuits. Thus, the major focus of the new intellectual tradition was to defend the Christian faith against opponents (eventually leading to scholasticism). For this purpose, the logical tools of Greek philosophy proved invaluable; for example, the use of the Platonic notions such as the immortality of the soul. Nevertheless, detractors also appeared who felt that the Greek tradition was in error, or were more attracted to the Aristotelian framework that was opposed to Christian doctrine at essential points. The Church, at times, supported philosophy (natural philosophy being a part). For example, Augustine (354-430) considered Greek philosophy useful. That is, natural philosophy (astronomy to regulate the religious calendar) could contribute to the proper interpretation of Christian doctrine. While space precludes a detailed discussion here, suffice it to say, that the Church was at times the patron of the sciences, but the practice of science was now chained to a rigid worldview that, increasingly, did not allow alternative views or new ideas.

3.4.1. The Islamic Contribution

In the Greek East, the consequences of the dissolution of the Roman Empire were not as severe. The capital of the Byzantine Empire, Constantinople, did not fall to Islam until 1205, in contrast to Rome in the 5th century. While the decline of natural philosophy and mathematics also declined, it was more gradual. Nevertheless, the fathers of the Greek Church also had the same aims as the Latin West.

In another part of the world, a movement was stirring that eventually had a major impact on natural philosophy in the West. The rise of Islam, while also religiously motivated, did not have quite the same rigid doctrinaire approach to epistemology as Christianity. The subordination of the Arabic Peninsula in 632 by Mohammed's followers marked the rise of Islam. In a little over a century (630-750), Islam had conquered an empire the size of Alexander's. As the Islamic rulers encountered educated Persians and Syrians, they were receptive to the spread of Hellenization. This process increased after 749. Moreover, by the start of the 8th century a number of Arabian scholars began the process of translation of Greek works into Arabic. One of these scholars was Hunayn ibn Ishoq (808-873) who deserves special mention. Assisted by his son, nephew and others, he translated the Greek, not literally word-for-word, but carefully to maintain meaning, into Arabic. As this translation process expanded with others, almost all the Greek mathematical, philosophical and medical corpus, was rendered into Arabic by 1000 AD. Presumably, Islamic scholars found these Greek epistemological works useful. Although scientific thought was never more than marginally acceptable within Islam, numerous original contributions were made (see Huff, 1993; McLeish, 1991). The scholar responsible for initially introducing the nine numerals (originally derived from the Indian subcontinent) and the zero that we currently use was Al-Khwarizmi (*ca.* 680-750). We also are indebted to Islam for *al-jabr*, or algebra and for *algorithmus*, algorithms, the latter arising from a Latinized name of an Arabic scholar (Struik, 1987). Other scholars of note include Al-Bottoni (d. 929) who introduced mathematical improvements into Ptolemaic astronomy, and Ibn al-Haytham an 11th century Muslim who made advances in astronomy (he criticized the Ptolemaic system as false), mathematics and optics. Particularly important was the work of 12th century Spain scholars such as Ibn Bojja (Avempace), Ibn Tufayl, and Ibn Rushd (Averroes). They tended to criticize the Ptolemaic system as physically impossible, although none anticipated the heliocentric theory. Only with Ibn al-Shatir (d. 1375) is there a suggestion of parallelism with Copernicus (Lindberg, 1992; Huff, 1993).

Space limits a detailed assessment of these rich and varied accomplishments. This Islamic intellectual movement lasted for 500 years until the 13th and 14th centuries, when a decline set in. Reasons for this are not entirely clear but explanations may lie in an increase of religious fervor that was less tolerant as seen in the burning of books in Cordoba. Also playing a part may be the general collapse that eventually occurred in the West as territories of the Islamic Empire fell into the hands of Christians, and in the East into the hand of the Mongols.

It is singularly fortunate that these products of Islamic science were not lost in these upheavals. Starting in the 10th century, a revival of learning appeared in the West, initially based on traditional Latin sources, it became apparent that a much richer source of knowledge was available from Islamic sources. The earliest introduction of the Arabic numerals and arithmetic was probably due to Gerbert of Aurillac (945-1003) who became Pope Sylvester II in 999. Based on a visit to Islamic Spain, Gerbert of Aurillac was able to absorb knowledge from Arabic sources (Ifrah, 2000). His teachings at the diocesan school at Reims did much to re-awaken interest in mathematics in the Latin West. He was the first to introduce Arabic numerals; however, he only brought back the first nine numbers and no zero. In spite of the teachings of his disciples at Cologne, Chartres, Reims and Paris, little impact was made on their contemporaries, if anything opposition toward anything 'Saracen' increased. It took a second introduction, initiated by Richard I (The Lionhearted) and the numerous crusades (1095-1270); which, as an unintended secondary consequence, re-introduced the arithmetic of the Indo-Arabic School (Ifrah, 2000).

By the 12th century, along with the creation of universities, Arabic translations became a major scholarly activity. One of these scholars was Gerard of Cremona (*ca.* 1114-1187) whose search for Ptolemy's *Almagest* led to Toledo. He learned Arabic and stayed over thirty years. His translated close to eighty books, an astonishing output. By the end of the 12th and the beginning of the 13th century, a majority of the surviving Greek and Arabic philosophical treatises had been recovered and translated in the Latin West.

3.4.2. Early Medieval Philosophical Developments

As scholars began to study the newly translated texts, they discovered their pagan character, which, in certain cases, led to potential difficulties, especially if they intersected with theology. Aristotelian concepts especially presented problems because they ran counter to the accepted worldview based on Platonic principles. Conceptual difficulties with Aristotle's views did not immediately arise at Oxford, but they did in Paris, and this led to a series of edicts (1210, 1213, 1231) forbidding instruction of Aristotle's natural philosophy. Nevertheless, by 1255, in a complete turnaround, scholars mandated that of all of Aristotle's works were to be part of the teaching curriculum. Thus, Aristotle became the cornerstone of scholasticism. Robert Grosseteste (1168-1253), the initial chancellor of the University of Oxford, was a driving force in this revival of learning. He was one of the first to deal with Aristotle's scientific method, which he viewed as both deductive and inductive. He attempted to characterize Aristotle's conception of *form* into quantitative terms (Crombie, 1961). His ideas influenced his pupil Roger Bacon (*ca.* 1220-1292), although not himself an experimentalist or mathematician Bacon made mathematical and experimental methods the keys to natural science. However, Bacon also passionately defended science in the service of the faith (Singer, 1959). Albertus Magnus or Albert the Great (*ca.* 1200-1280) was the first to offer a comprehensive interpretation of Aristotle's philosophy, which he regarded as necessary preparation for theological studies. It is not an exaggeration to suggest that the acceptance of Aristotelianism in the 13th century was due in large part to him. His student Thomas Aquinas (*ca.* 1224-1274) continued the work of his mentor of merging Aristotelian thought with Christian doctrine. This merging was never without controversy and led to conflict between conservatives and progressive views within the Church. In particular, one opponent was Siger of Brabant (*ca.* 1240-1284). Siger advocated the total separation of philosophy from theology. A position he later carefully 'recanted'. Nevertheless, the position of Siger and his followers led the conservative faction to issue the edicts of 1270 and 1277 specifically forbidding the teaching of certain Aristotelian principles on the pain of excommunication (Lindberg, 1992).

In spite of these restraints, epistemological sophistication was steadily increasing leading to skepticism. That is, whether philosophical arguments could even be used to directly address theological concerns. John Duns Scotus (*ca.* 1266-1308) and William of Ockham (*ca.* 1285-1347) both questioned the ability of philosophy to demonstrate articles of faith with certainty. Thus, articles of faith were not open to philosophical examination; but had to be accepted by faith alone. A view still espoused by theologically-minded existentialists such as Søren Kierkegaard (1938) among others. This eventually led to an uneasy peace between natural philosophy and theology. Each of which, was now beginning to be relegated to separate spheres of influence. Leaving these interesting philosophical issues aside, we now turn to other more relevant developments that eventually led to modern science.

3.4.3 The Ptolemaic Worldview

This section ends with the accepted cosmological view of the universe as shaped by Aristotelian thought and legitimized with the Ptolemaic mathematical model. It was comfortably ethnocentric and placed the earth at the center of the universe. The earth, at the center, consisted of the terrestrial region composed of the spheres: earth, water air and fire. Above that where other spheres radiating outward starting with the moon, Mercury, Venus, the sun, Mars, Jupiter and Saturn, and beyond them lay the fixed stars (Hall, 1994). Observations suggested that these planets traveled with uniform circular motions. The fact, that in reality, things were not so neat had been recognized (refer to Section 3.4.1. for the Islamic skepticism).

The problem of retrograde motion, for example, required adjustments. The Ptolemaic model consisting of eccentrics and epicycles was developed to specifically answer these observational discrepancies. That the model was successful can be attested to the fact that it was scarcely questioned for over thirteen centuries. Moreover, because the earth was the center of the universe in the Ptolemaic system, this conveniently served Christian doctrine. So what factors led to the breakdown of the Ptolemaic system? Numerous reasons have been offered and these will be examined below where we discuss the developments that led to the Renaissance and with it, the rise of modern science.

3.5. FROM THE RENAISSANCE TO THE ENLIGHTENMENT

3.5.1. The Copernican Revolution

The first phase of the transition toward a mechanistic view of the world occurred during the Renaissance (1450-1550). This was a period of admiration and assimilation of ancient Greek and Roman ideals, as well as a period of unparalleled experimentation in many fields, made possible by greatly improved economic conditions, initiated with the beginning of the breakdown of feudalism leading to the rise of capitalism based on trade (Bernal, 1971b). More importantly, one essential aspect of the ecclesiastical edifice of the Church; namely, the cornerstone of scholasticism, Aristotle, had come increasingly into question.

The epitome of the re-discovered humanistic spirit of the Renaissance can be seen in the works of three of its greatest minds. The first one of these was Copernicus whose *De Revolutionibus Orbium Coelestium* defined the heliocentric universe. We will take up his work in some detail below. The second one was Andreas Vesalius who's *De Humani Corporis Fabrica*, was the first complete anatomical study of the human body. Both of these books were published in the same year, 1543. These two books, as well as a host of others, did much to break down the longstanding tradition of medievalism and scholasticism that had characterized known knowledge for more than a millennium. The third individual, whose contributions are so well known that they scarcely need to be mentioned, was Leonardo da Vinci (1452-1519).

With the Renaissance came, first and foremost, observation and description, the two central aspects of modern science, Leonardo da Vinci's anatomical studies are two good examples. Nevertheless, while there was now recognition of the human body as a complex machine, and the terms of physiology and pathology, so central to modern medicine, had been coined by Jean Fernel (1497-1558), little substantial had changed. That is, the classical picture derived from Galen (130-200 A.D.) remained intact (Bernal 1971b). Moreover, the recognition of the complexity of the human body, did not, in it self, reveal anything about process, function or evolutionary relationships. These had to wait for another two centuries of scientific developments. Nevertheless, the classic authoritarian worldview held for last two millennia was beginning to crack, and be replaced by direct observation and experiment.

Leonardo da Vinci typifies the Renaissance with all its spectacular successes and failures. Art had become scientific so to speak. At no other time in history have the visual arts had such an effect on science (Bernal 1971b). This is especially apparent in the invention of *perspective,* which was an endeavor to depict visual elements as they really appeared. It is also seen in anatomy with the accurate portrayal of the human form. Nevertheless, while neither approach was quantitative yet, it foreshadows morphometrics. Leonardo attempted to state this relationship between art and science as:

> The science of painting deals with all the colors of the surfaces of bodies and with the shapes of the bodies thus enclosed; with their relative nearness and distance; with the degrees of diminutions required as distances gradually increase; moreover, this science is the mother of perspective, that is of the science of visual rays. (Quote from Bernal (1971b:390) of Leonardo's *Paragone* text).

While Leonardo's numerous contributions are undoubted, his drawings of such weapons as tanks and his models of flying machines, etc., never could have worked because of the absence of needed methods of propulsion and the lack of a quantitative knowledge of statics, aerodynamics, etc. He wanted to understand the underlying nature of what he saw, hence his studies of optics, anatomy, animals, plants, etc. His accomplishments were based on a fertile imagination that could inspire others rather than on a systemic approach or mathematical skills. However, it is especially important to remember that Leonardo was, above of all, a painter and his various interests were often more focused on the need to increase his skills in the area of painting (Costantino and Reid, 1991).

Nicholas Copernicus (1473-1543) has been considered the father of modern science (Neyman, 1974) although he depended on the work of others. Copernicus' famous work *De Revolutionibus* was published in the year of his death. However, it was preceded by a work entitled *Narratio Prima* written by an admirer, a Lutheran astronomy professor, Georg Rheticus (1514-1576), in 1540. Although, it has been inferred that Copernicus acknowledged his debt to Aristarchus of Samos for the heliocentric system no mention of it can be found in *De Revolutionibus.* However, Aristarchus is mentioned in an earlier manuscript (Crowe, 1990). Nevertheless, the Copernican system was still based on the idea of uniform circular motion, which required an acceptance of the epicyclical motions. Thus, his model of the universe is only a moderate improvement over that of Ptolemy as far as the observations of the planetary motions are concerned. In fact, the Copernican system used thirty-four circles, in a similar way to the Ptolemaic system, making them equivalent in predictive accuracy (Kuhn, 1957; Singer, 1959). Ironically, when judged on observational criteria, the Copernican system was neither more accurate nor simpler than the Ptolemaic one. This was in spite of the preface of the *De Revolutionibus* where Copernicus touted his cosmology as being superior to Ptolemy. So why was the Copernican system a success?

The first reason was that by proposing a heliocentric universe instead of geocentric one, Copernicus provided a model, which eventually explained rather diverse observational phenomena. The second reason was his scientific outlook and his willingness to abandon dogma backed by tradition, not by evidence (Neyman, 1974). He insisted on conformity of theory with observation, a procedure diametrically opposed to Church doctrine. That is, if observation disagreed with doctrine it was often conveniently ignored. [24] Nevertheless, it took another century before Church dogma became modified in accord with astronomical theory. By accepting a heliocentric universe, Copernicus could now qualitatively explain two of the objections leveled against the Ptolemaic system. One of these was that a correct explanation; that is, one in accord with observation, was now possible for retrograde motion. The other one was that the varying brightness of the planets could also be explained. Thus, a single change, replacement of the earth with the sun, explained a rather diverse set of observations. For the next generation of astronomers, the starting point was the heliocentric approach and the earth's motion on which they now focused. For laymen, the *De Revolutionibus* was largely unreadable because of its complexity and use of mathematics. Most popular works either did not mention Copernicus at all or dismissed him in a sentence or two, until the start of the 17th century (Kuhn, 1957). This slow assimilation (astronomers being the one exception) also prevented the large-scale lay and clerical opposition so pronounced in the next century. Nevertheless, the Copernican system eventually was to influence thought beyond the bounds of observational astronomy. The relatively simple change of a geocentric to a heliocentric conception of the universe carried widespread and disturbing implications. It led to a fundamental shift in worldview, a paradigm shift initiated by Copernicus but gradually completed by others. By the start of the 17th century, Catholic and Protestant doctrines had hardened with respect to Copernicanism, its teaching being forbidden in 1616 and again in 1633. However, even before that, the case of Giordano Bruno (1547-1600) brought widespread attention to the Copernican system. Bruno was burned at the stake, not for supposedly holding the dangerous Copernican beliefs (which he did), but for his view of the Trinity. That being a heresy for which other Catholics had been executed earlier (Kuhn, 1957).

After Copernicus, the most influential astronomer was the Dane Tycho Brahe (1546-1601) who proposed a different cosmological system. In 1572-74, he noticed the birth and death of a new star (what today would be called a nova) of which he wrote a short work (Crowe, 1990). Such an observation raised questions with the Aristotelian concept of an immutable, unchanging universe. In 1574, he began to give lectures at the University of Copenhagen and in 1576 laid the cornerstone for an observatory called Uraniborg. He began a series of astronomical observations there that resulted in a considerable increase in accuracy (in the absence of telescopes). According to Kuhn, his accuracy was double that of earlier observers; that is, his planetary positions were measured to an accuracy of 4 minutes of arc. In 1577, he produced a third system of planetary motions rivaling those of Ptolemy and Copernicus. This system, while geocentric, was simpler in that it did not require epicycles. It retained the basic principles of the Ptolemaic approach such as being geocentric. Brahe was and remained anti-Copernican in his views and his system was intended as a compromise. This system, viewed today as a curiosity, but coupled with the much more accurate observations of Brahe, provided the basis for Johannes Kepler (1571-1630) who was in Brahe's employ

---

[24] It should be noted that the *De Revolutionibus* was placed from 1616 to 1822 on the Index of books forbidden to be read by Catholics.

for the last few years of Brahe's life in Prague. Brahe had hired Kepler to put his cosmological system on a more solid mathematical footing (Crowe, 1990).

In a number of substantial ways, Kepler's work can be said to overshadow that of Copernicus. With Kepler, we finally have arrived at a correct mathematical model. Kepler had started out endorsing the Copernican system in an early publication *Mysterium cosmographicum* (1596) in which he used the Platonic solids to define planetary orbits. Although eventually increasingly critical of the mathematical underpinnings of the Copernican system, he remained a life-long Copernican. His first rejection of the quantitative basis of the Copernican/Ptolemaic universe based on uniform circular motion occurred when he examined the orbit of Mars. Mars had the largest epicycle to account for its retrograde motion. This took ten years, since he had to account for both the motions of earth and Mars. He experimented with various mathematical approaches using circles to account for of the motions of Mars as observed by Brahe. After his unsuccessful use of circles, he started using ovals and eventually stumbled on the use of ellipses. Then allowing for a relaxation of the traditionally held view of uniform motion, he was able to develop a mathematical model that, for the first time, correctly predicted the orbit of Mars. Thus, with a single geometric curve, Kepler was able to dispense with the morass of *ad-hoc* artifacts such as epicycles, etc. (Kuhn, 1957). His results were published in 1609 in his *Astronomia nova* in which he first presented what are now known as Kepler's first and second laws of motion. One is a consequence of the other. He found that Mars: [1] moved in an elliptical orbit with the sun at one of its foci and [2] that during the rotation of Mars around the sun, it sweeps out equal areas in equal times. Kepler's enduring claim to fame is not only as the author of the correct mathematical model of planetary motion, but for his scrupulous attempts to fit his analytical model to observation. He represents an early example of the application of the modern scientific method. In a sense, Kepler's relaxation of uniform motion is as profound an idea as Einstein's idea of the equivalence of mass and energy. Although Kepler's work was received with mixed feelings at the time of its publication, he clearly represents a continuation of the break with the traditional worldview based on Aristotle and subsequently accepted as Church dogma. Nevertheless, Kepler's work and with it the Copernican worldview did not gain wide acceptance until the end of the 17th century.

The post-Renaissance period (1550-1650) represents a period of continuous consolidation and gradual acceptance, in spite of considerable clerical opposition, of the new worldview. According to Bernal:

> In science, the period includes the first great triumphs of the new observational, experimental approach. It opens fresh from the first exposition of the solar system by Copernicus and closes with its firm establishment—despite the condemnation of the Church—through the work of Galileo. It includes in its scope Gilbert's description in 1600 of the earth as a magnet and Harvey's discovery of the circulation of blood. It witnesses the first use of the two great extenders of visible Nature, the telescope and the microscope (Bernal, 1971b:411).

Galileo Galilei (1564-1642) contributions particularly stand out; he is, in a sense, the first scientist in a modern mold. He understood like few others, that observation of phenomena was not sufficient in itself, that it was also necessary to inquire how such a system could exist. While initially trained along scholastic, that is, along Aristotelian lines, by 1590 he had already developed objections to the physics of Aristotle (Singer, 1959). Galileo succeeded, where others such Leonardo da Vinci had failed, by formulating the motion of bodies (study of dynamics) in mathematical terms. He represents the real beginning of experimental physics. For example, he suggested that the path of projectiles was based on compound motions. That is, composed of two motions, one uniform and the other accelerated, which anticipated Newton's First Law (Singer, 1959). However, while much closer to the truth, he also failed in his mathematical application of the velocity of falling bodies (he presumed that speed was gained in proportion to distance). Had Galileo been able to create tables of falling distances as a function of time, he might have stumbled on the correct explanation. However, he did not. He did not grasp the rather difficult concept of velocity as a function of distance over time; that is, the idea of a differential, dx/dt. This had to wait for Newton and Leibniz. Nevertheless, he was able to use these ideas even if he was not able to formulate them. He combined careful experiments with mathematics and thereby provided the first clear example of the methods of modern physics (Bernal, 1971b). By 1636, he is clearly occupied with the ideas of force and motion and their interrelationships, ideas that will re-appear in the work of Descartes and in a more sophisticated form in the *Principia* by Newton some fifty years later (Singer, 1959). Galileo had considered *extension and movement* as primary physical realities. By extension, he meant length, width and breadth of form (*i.e.*, matter). For Galileo primary qualities were *static* (forces in equilibrium) and *dynamic* (out of equilibrium). From these notions, he attempted to lay down the laws of motion. As Bernal observed:

> Galileo stated more clearly than anyone before him that the necessary and intrinsic properties of matter—the only ones in fact that could be dealt with mathematically, and therefore with any certainty—were extension, position and density. All others, 'tastes, smells and colours, in regard to the objects in which they appear are nothing more than mere names (Bernal, 1971b:433).

Galileo's work here hints that he was, as Leonardo before him, preoccupied with the concept of form (matter). The division of into primary qualities (those to be dealt with mathematically) and secondary ones such as 'colors', etc., will reappear in the work of Descartes. Again, these ideas anticipate modern morphometrics.

In 1608-09, the telescope was invented and Galileo was one of the first to make major use of it. In a few nights of observation, of the heavens he saw enough to shatter the long-held Aristotelean system (Bernal, 1971b). Interestingly, in a letter much earlier to Kepler (1597), he had already confided that he had accepted Copernicanism. Galileo discovered new stars, mountains and craters on the moon, the moons of Jupiter, phases of Venus, etc., all observations that changed earlier conceptions of the heavens and the bodies in it. After Galileo came amateur astronomers as the telescope had become widely available. However, with the heightened interest in observational astronomy, there also came an increased opposition from the holders of the traditional viewpoint. The opposition of the Church toward the telescopic observations of Galileo and others was symptomatic of the increased opposition to Copernicanism in general. One has to remember that that a complete system of philosophy had been developed over centuries, a scheme that included the moral or spiritual realm with the physical realms composed of the earth and the heavens. This was a system that satisfied the medieval mind. Thus, the fact that Galileo had rendered a hole in the fabric of the old worldview derived from Aristotle initially, did not, have much effect. If the Church had recognized the implications of the mechanistic conception of the world, their negative reaction would have been swift. It was providential for Galileo that they initially did not. Abandonment of Aristotelean logic implied abandonment of much religious teaching (Singer, 1959). It also reflected the deep-seated fear that the traditional Ptolemaic worldview, held for centuries, would be overturned resulting in moral and spiritual chaos.

Marshaling all the observational facts now at his disposal, Galileo set out to demolish the Aristotelian edifice. In the process, he finally incurred the wrath of the Church. His polemical book, *Dialogue Concerning the two Chief Systems of the World, the Ptolemaic and the Copernican*, written in Italian to gain a larger audience, came out in 1632 and was dedicated to the Pope. It led directly to the famous trial. Because Galileo was 70, had an international reputation as a scientist, was an undoubted Catholic and had influential friends in high places, the sentence meted out was comparatively lenient. He was initially sentenced to life in prison, but that sentence was quickly reduced to house arrest. Nevertheless, the sentence marked the zenith of the conflict between science and religious dogma. The verdict was poorly received outside of Italy and the controversy was eventually quietly dropped (Bernal, 1971b). Nevertheless, Galileo's works, as Copernicus' before him, ended up on the Index, that list compiled by the Church of works considered injurious to the faithful.

Galileo, as Bruno before him, had the unfortunate misfortune to be caught in the reactionary movement that arose against the new developing worldview. While this change in worldview did not lead to a rejection of religion per-se, it did begin to alter the relationship between individuals and God. It made the relationship more personal and the role of the Church less central. The Church hierarchy felt increasingly threatened by this new worldview and the backlash, alluded to earlier, became inevitable. It led to a sustained campaign by the Church to suppress, by any means possible, the adherents of this new view. On another front, the Church also faced the increasing moral corruption of the Church hierarchy. This eventually led to Martin Luther (1483-1546) and the Reformation (1517-1535) with the consequent rise of Protestantism.

Aside from the above schism that led to Protestantism, the repercussions of this change in worldview continued well into the 18th and 19th centuries and are still being felt today. It is ironic that the rise of the creationism movement in the United States in the last decades of the 20th century is also a testament to the enduring longevity of that ancient and discredited worldview. This change in worldview continues to have perplexing psychological effects. One reason for this state of affairs is existential. That is, the acceptance of this new worldview affects the very core of human existence because it displaces the earlier, more psychologically appealing, worldview. That earlier worldview had emphasized a much more psychologically comfortable relationship of man within the world and the universe. That is, man, as a creation made in God's image, was central to the universe as seen by the position of the earth, vis-à-vis the Ptolemaic system. Viewed in this light, the centuries-long opposition and still, at times, lingering hostility of the Christian Church toward all scientific endeavors, while perhaps not justified, becomes more understandable.

3.5.2. Developments Leading to Newton and Beyond

By the beginning of the 17th century, a small group of exceptionally able individuals such as Copernicus and Galileo were able to lay down what finally became the correct foundations of astronomy as outlined in Section 3.5. One of the factors that led to the wide acceptance of astronomical principles was the practical knowledge it provided for such endeavors as navigation. Navigation had become increasingly essential to the voyages of exploration, which characterized the 15th to 17th centuries. More importantly however, it also laid the groundwork for an emphasis on the study of natural phenomena, which required new experimental and mathematical methods to analyze the problems that now arose (Bernal, 1971b). This shift in attitude and approach to the natural world, beginning in the 17th century and ending with the close of that century, while subtle was a scientific revolution. As Bernal has indicated:

> It amounted to a *scientific revolution*, in which the whole edifice of intellectual assumptions inherited from the Greeks and canonized by Islamic and Christian theologians alike was overthrown and a radically new system put in its place. A new quantitative, atomic, infinitely extended, and secular world-picture took the place of the old qualitative, continuous, limited and religious world-picture which the Muslim and Christian schoolmen had inherited from the Greeks. The hierarchical universe of Aristotle gave way before the world machine of Newton (Bernal, 1971b:375).

This shift represented a profound change in worldview. A shift from a medieval and theologically dominated view of an idealized world, biblically created, to one that was controlled by the forces of nature and which operated according to predictable mathematical laws. Concomitantly, this also led to a change in the acquisition of knowledge from one based on the authority of Greek and Roman literary sources, to one founded on experience and grounded in nature. Put somewhat differently, the emphasis on the pursuit of knowledge had subtly changed. It had changed from the earlier one that edified God to one that had discovered nature and the physical world.

While increasing acceptance of the Copernican-Kepler system was found among astronomers, and opposition from the Church beginning to wane, problems remained. One of these was the need for some kind a theoretical basis for the observed astronomical phenomena. Science was still overwhelmingly descriptive and any kind of explanation in physical terms was still lacking. A synthesis of some sort was needed. One physical clue was the work of William Gilbert (1546-1603), physician to Queen Elizabeth I, whose experimental study of magnetism was published in Latin in 1600 and known by its shortened title as *De Magnete*. This book is an important contribution that stands on its own merits, as well as influencing the subsequent work of Newton. Gilbert's imaginative ideas led him to suggest that the earth was a giant magnet and it was magnetic attraction that accounted for planetary motion. This is the first plausible explanation, one in purely physical terms rather than metaphysical ones, of the ordering of the heavenly bodies (Bernal 1971b; Singer, 1959).

Another step forward was the increasing recognition that mathematics had a major role to play in science as seen in the accomplishments of Kepler. More developments that were fundamental appeared with Francis Bacon (1561-1626) and Rene Descartes (1596-1650). While both men influenced the development of philosophy, science and the scientific method, they also reflected the medievalism of the period. Both approached science in contrasting ways. Bacon's emphasis was *inductive* (as Aristotle before him), collecting materials, carrying out experiments and drawing conclusions from the physical observations.[25] Bacon's empiricism placed emphasis on the experimental character of scientific methods and his method of induction influenced many of the founders of the Royal Society who viewed themselves as his disciples (Hall, 1981). Bacon belonged to the tradition of the encyclopedists and his concept of the organization of facts led directly to the formation of the Royal Society of England. Bacon however, had no sympathy with mathematical ideas; he was, first and foremost, in the naturalist tradition (Bernal, 1971b). He was much less of a theorist than Descartes, much more of a collector of facts. He thought that results would become self evident once all the facts had been collected (Singer, 1959). Thus, Bacon actually accomplished very little. He was anti-Copernican in outlook and did not recognize or appreciate that a fundamental change in worldview was occurring.

Descartes on the other hand, used a largely *deductive* approach.[26] Descartes attempted to explain the physical observations by carefully applying a set of analytical rules based on rational thought. His work made a major break with the past in that he proposed a system that could be used to describe the material world in strictly mathematical terms. His analytic method appeared in his 1637 essay *La Geometrie*, which connected the idea of motion to a geometric field (Singer, 1959). The Cartesian method of co-ordinate or analytical geometry, joined Greek geometry with Arab algebra, and can be considered a mathematical achievement of the first rank. It provided a method by which; *e.g.*, a curve in 2-space could be completely described by an equation. It was to subsequently affect the direction of mathematics for the next two centuries. The fundamental contribution of Descartes is that he connected number with form and in this sense can be considered as one of the forerunners of morphometrics. While still largely Platonic in character, his approach, nevertheless, anticipates modern developments and his work has a direct bearing to morphometrics as will be demonstrated in later chapters.

Another one of Descartes' numerous contributions was his ambitious attempt to provide a system of how the universe was constructed. His universe was composed of material particles arranged in vortices, called a *plenum*. He believed that the sun was at the center of one of these vortices and the planets orbited around it.

A fact not always recognized is that Descartes' co-ordinate method, so essential to modern mathematics, was just a part of the Descartes' system of metaphysics, his attempt to explain the primary cause of the natural world—God. In this sense, he is strictly in the Aristotelean-medieval tradition of scholasticism. With scholasticism, knowledge was defined as coming from faith or revelation. This was based on careful

---

[25] In simple terms, *induction* refers to the logical process of drawing a general conclusion from specific premises or observations. Inductive inference can also be defined as arguing from the 'specific to the general'. It has to be viewed in probabilistic terms (see Reichenbach, 1951). However, reference should also be made to the *hypothetico-deductive* method proposed by some philosophical opponents of induction, especially in terms of verification (see *e.g.*, Popper, 1959, and elsewhere).

[26] Again, in simple terms, *deduction* refers to a logical procedure that derives a particular conclusion from general premises (often *a-priori*). Here one argues from the 'general to the specific'. Certain philosophical schools have appeared such as Logical Positivism that have put considerable weight on terms such as deduction (Ayer, 1959). However, considerable controversy continues to surrounds induction and deduction, and the reader should consult the relevant philosophical texts for more complete definitions.

reading and reconciling of the scriptures with the logic of the Greeks in an attempt to justify Christian faith with reasoned arguments. This became the accepted medieval view, although as mentioned earlier, questioned by such scholastic philosophers as William of Ockham (*ca.* 1285-1347) and Roger Bacon (*ca.* 1214-1294).

Descartes, as others, viewed the division of knowledge into physics (the external world) and metaphysics (moral world). The material of physics, matter, is to be explained in terms of their sizes and shapes (form), and their movement since these were observable phenomena (and for him mathematically describable), the so-called primary qualities already encountered with Galileo. Secondary qualities like color, temperature, smell, etc., are subjective (and more difficult to quantitatively describe). Thus, for Descartes, as Galileo before him, only size and shape (qualities of extension) and motion are intrinsically possessed by all matter. In modern morphometric terms, this distinction of primary and secondary qualities, is more apparent than real. Size, shape, color, temperature, surface texture, and a host of other qualities can *all* be considered as aspects of form (matter) making the above distinction into primary and secondary qualities quite unnecessary, although computational problems still remain (Lestrel, 1997). In addition to the two 'qualities' mentioned here, there was a third one, which lay outside the provenience of physics. This third set of qualities had to do with the passions, love, faith, etc., *i.e.,* moral qualities or metaphysical qualities beyond measurement and hence, beyond physics. Descartes achievement lies in recognizing the importance of knowledge derived from the physical world (matter) as primary. Knowledge of which becomes potentially measurable by his co-ordinate system. This is in dramatic contrast with the prevailing view that still held that all knowledge was based on faith or revelation. For Descartes separation of knowledge into physics and metaphysics:

> ... is a logical consequence of his reduction of sensory experience first to mechanics and then to geometry (Bernal, 1971b:445).

Thus, for Descartes only measurable entities (the primary qualities) of matter are the basis of science and the third set of qualities mentioned above, belong to the realms of faith and revelation. However, having separated these two spheres of knowledge and placing an emphasis on the first rather than the third put Descartes directly into conflict with the ecclesiastical thought of the Church. The problem that now arose had to do with the increasingly mechanistic view of physics, or more broadly, natural philosophy. If all matter, including man and animals, can be reduced to primary qualities such as extension and motion, a view that Descartes subscribed to, then how can one also account for the spirit or rational soul—the *élan vital* of Henri Bergson (1859-1941)?

Descartes was able to cleverly connect as well as separate the two spheres of knowledge, physics and metaphysics, thus directly avoiding the religious issues that had led to the burning at the stake for Michael Servetus (*ca.* 1511-1553) in Geneva, Bruno (1548-1600) in Rome, and one Lucillo Vanini in 1619 in Toulouse for atheism. In 1624 disputations against the work of Aristotle were prevented by official edict in Paris and at the request of the Sorbonne, a ban was placed in the teaching of any proposition critical of any of the ancient authorities (Sorell, 1987). Having arrived in Paris in 1626 where he was to stay for two years, he was undoubtedly influenced by the atmosphere present there. Thus, in his *Discourses* Descartes argues for the existence of God. As Sorell puts it:

Descartes' reasoning starts with the discovery of a necessary connection between his thinking and his existing, moves to the necessary objectivity of his idea of God, and finally reaches conclusions about the intellectual capacities of creatures who can trust a benevolent God not to place false simple thoughts before their minds (Sorell, 1987:52).

Therefore, this first given, that I am thinking means that I exist, led to the famous phrase, *Je pense donc je suis*—"I think, therefore I am". Descartes here is indicating that by thinking he is conscious of self. This became the rationale for the existence of God as a perfect being. For Descartes, the ability to think of a perfect being, implies a given that God exits. The Platonic nature of all this is quite evident. From this first principle, Descartes argued that other metaphysical truths could then be deduced. As Sorell indicted:

For what he took to be the subject matter of metaphysics in general, we have to refer to the preface of the French edition of the *Principles of Philosophy*. 'The first part of philosophy', the preface says, 'is metaphysics, which contains the principles of knowledge, including the principal attributes of God, the non-material nature of our souls and all the clear and distinct notions that are in us' (Sorell, 1987:56).

Here Descartes is placing certainty upon thought or intuition, and thereby independent of experience. He held that the mind could attain universal truths. For Descartes the starting point of all human knowledge had to be *a-priori*.

Thus, in an attempt to answer the question raised earlier — the need to account for the spiritual nature of man in a increasingly mechanical universe, Descartes presented what has been viewed as a rather naïve explanation, however quite in keeping with the medievalism still prevalent in the 17th century. He attempted to identify where the physical and metaphysical natures of man intersected, the old and enduring mind-body problem. This distinction between the mind and the body as visualized by Descartes has become known as Cartesian Dualism (Sorell, 1987). When pressed for clarification of his mind-body theory, he relied on three concepts, which were:

... As regards body in particular, we have only the notion of extension...and as regards the soul as such we have only the notion of thought...Finally, as regards the soul and body together, we have only the notion of their union (quoted from Clarke, 1982:26).

For Descartes the seat of the soul could be found in the pineal gland located in the cranium. Once considered to be a vestigial sensory organ, but having now no apparent function, it was a reasonable candidate if not for the seat then at least for the point of departure of the rational soul.

While the logic that Descartes developed here may seem peculiar in 20th century terms, it assured Descartes safety from prosecution especially given the increasing protests about his work that had arisen from the universities. Finally, what is perhaps an oversimplification of Descartes philosophical views, which were initially largely deterministic in nature as can be seen from the *Discours de la Methode* (1637), only later to increasingly embrace, possibly because of ecclesiastical pressures, a more metaphysical outlook as seen in his *Meditations* (1641). Nevertheless, the substantial accomplishments of Descartes were to be beneficial to science, physiology and philosophy.

The year 1642 is memorable for the fact that Galileo died that year and Issac Newton was born. Thus, it is by middle of the 17th century that the beginnings of what was to become modern science can be finally glimpsed. Beginning with Copernicus and Kepler, leading to Galileo, observation and description had become legitimate scientific endeavors, and now with Newton, observation became coupled with explanation and the synthesis was finally completed.

Thus, the greatest triumph of the 17th century was the development of a system that was capable of accounting for the motion of the earth against the stars. This interest in the heavens, once mostly philosophical in outlook, leading to a replacement of Aristotle's cosmology, now became increasingly practical. What was now needed was a system that allowed prediction so that more accurate astronomical tables could be created. Especially acute was the need for the identification of longitude by mariners. In practical terms, this reduced to a matter of time and led to Greenwich time—from which local time (longitude) any place at sea could be determined with the use of a chronometer. However, within astronomy a much more fundamental problem had appeared. If the Copernican-Kepler theory was to be held as valid, as observations seemed to now suggest, then what was the mechanism that accounted for the movements of the solar system? Descartes had earlier postulated a system of vortices to explain this and his system had become widely accepted. However, this system was vague and non-quantitative. Pierre Gassendi (1592-1655) placed renewed emphasis on the older Greek view of atomism championed by Democritus among others, the so-called corpuscular hypothesis. The corpuscular hypothesis was to effectively supplant Descartes' system. Its advantage was it facilitated the application of a mathematical approach based on point-like particles rather than larger units of matter having extension in space, which would be much more difficult to handle quantitatively.[27] Gilbert had also provided a partial answer with the idea of attraction (magnetism), but it took an exceptional individual with unsurpassed mathematical ability to find a decisive solution. This person was Issac Newton (1642-1727).

We are indebted to Edmund Halley (1656-1742), whose comet bears his name, for convincing Newton to work on the problem of gravitation. With the corpuscular hypothesis in hand, Newton simplified the problem of weight by concentrating it at the center (point-mass). Newton, basing his conceptual approach on laws, was able to conjecture that gravitation, the attractive power of the earth toward any body (*e.g.,* the moon) decreases as the square of the distance between them. Rather late in life, he published the inverse square law of gravitation in *De Philosophiae Naturalis Principia Mathematica* (1687) or *Principia,* as it has since become known. He had originally conceived these ideas in 1665 getting his cue from the work of Galileo, but did not see fit to publish at the time. In fact, Newton as late as 1679 still held to Descartes' system of attraction of planetary bodies as due:

---

[27] This is still largely true in morphometric terms, in the sense that it is quite difficult to completely describe in quantitative terms any but the simplest of biological forms. Consider trying to describe such a form as *e.g.,* the human cranium. This would not only, require measurements of size (or to use the medieval concept of extension - length, width and breath), but also shape, color, texture, etc. While these are ancient concepts, the issues raised then are still very much with us. See especially Chapters 5, 6 and 9.

... some secret principle of unsociableness of the ethers of their vortices (direct quote in Bernal, 1971b:480).

Newton's contribution lay in finding the mathematical method for converting physical principles into quantitatively calculable results confirmable by observation, and conversely, arriving at the physical principles from such observations (Bernal, 1971b).

The approach that he used was the infinitesimal calculus or as he called it, the *method of fluxions*. While others have had similar ideas, and Gottfried Wilhelm Leibniz (1646-1716) the great German mathematician is credited with independently developing the calculus, Newton deserves credit for applying it to solve critical problems in physics. Newton's *Principia* established the laws of motion for the planets. It also represents the first attempt to use a systems approach (Section 5.1.2). It is unparalleled in the history of science. In importance, it ranks equal to Euclid's *Elements*, Darwin's *Origin of the Species* or Einstein's concept of Relativity. In the *Principia* Newton not only demonstrated how universal gravity maintained the solar system, but he also presented it in quantitative terms rather than the more usual philosophic approach that had characterized cosmologies from the time of Aristotle to Descartes. With Newton, a fundamental shift took place, which was to have profound repercussions for the future of science. While elements of this shift can be traced in the work of Galileo, Kepler and others, Newton was most instrumental in setting the stage for modern science. Earlier investigators focused on *describing* the motions of the heavens, while Newton was the first one to attempt to provide an *explanation* of these heavenly movements (Singer, 1959). While both aspects are necessary, the distinction between description and explanation is central to modern science. Moreover, explanation leads to prediction in science.

It is with Newton that science and the scientific method finally matures and ushers in an unstoppable flood of developments in the search of new knowledge that has led to the 20th century and has continued unabated. By the last decades of the 17th century, the scientific method was beginning to be firmly established as the premier method for attaining knowledge. One reason for the success of science was that solutions to practical problems such as navigation, gunnery, etc., increasingly required the services of mechanics, astronomy and mathematics. This pattern has accelerated in the 19th and especially the 20th century. Another major factor was the foundation of both the *Royal Society of London* (*ca.* 1662) in England and the *Académie Royal des Sciences* (*ca.* 1666) in France. One of the reasons for the acceptance of the Royal Society was that there was a formal attempt to limit scientific discussion to legitimate scientific endeavors and to exclude metaphysical ones (Bernal, 1971b). Interestingly though, by the 1690's these scientific societies were in decline and to be 're-instated' in the 18th century as the industrial revolution gained steam. The foundations of the scientific method were now largely secure and prospering. Scientific endeavors had now been largely separated from metaphysics. Science, or rather natural philosophy as it was then called, became respectable and acceptable by the State if not by the Church. Nevertheless, the hold of religion remained considerable limiting much deviation from the general biblical scheme of creation and salvation accepted by Catholics and Protestants alike (Bernal 1971b). While the scientific seeds had been sown in the 17th century, it was to take another 200 years before:

... the day of theological domination over science had ended. It could still distort and delay the advance of science, but could not stop it. Religion was confined to the moral and spiritual sphere (Bernal, 1971b:497).

Before we return to the central focus of the book; that is, morphometrics, we need to first continue to trace the expansion of science into new fields of knowledge; namely the biological sciences. The development and application of quantitative methods in biology is the topic of the next chapter.

KEY POINTS OF THE CHAPTER

This chapter has briefly focused on the beginnings of science and the scientific method. Within that structure, the beginnings of the science of measurement can also be discerned. The chapter started with the very beginnings of measurement and ended with a sophisticated mathematical view of the universe. This was set within the framework of the history of science starting with early man and leading to the Ptolemaic cosmology. This worldview originating with Aristotle was accepted in Western Europe for over thirteen centuries. It was gradually replaced with the heliocentric system of Copernicus. The new cosmology, based on the Copernican system, eventually overthrew the Aristotelian one because it was grounded both in observation and in prediction. The predictive mathematical system was initially developed by Kepler and then refined into the modern form by Newton.

CHECK YOUR UNDERSTANDING

1. How does the development of language contribute to the early beginnings of measurement?

2. Trace the development of numeric systems from the beginnings of early man to the rise of civilizations.

3. Why are the Pre-Socratic philosophers so important for understanding the beginnings of science and scientific measurement?

4. It has been argued that Plato's philosophy was hostile to scientific interests. Comment on this thought in some detail.

5. Why did Aristotle, who was a student of Plato, disagree with his teacher? How did this disagreement affect science?

6. Trace the beginnings of theoretical cosmology from Thales to Ptolemy. What do these differing views offer about science and the scientific method?

7. Compare the cosmology of Ptolemy with that of Copernicus. How does each view attempt to reconcile observation with explanation?

8. Besides the position of the earth in the universe, what other significant differences are present between the Ptolemaic and Copernican worldviews?

9. Why was the Aristotelian view accepted for over a thousand years and why was the Copernican worldview so difficult to accept when it appeared?

10. What was Newton's contribution to cosmology, and why was it so important?

11. Consider the worldview that existed at the time of Newton or a little after, and compare it with that of Einstein. What conclusions can you draw about the scientific method then and now?

12. Trace the idea of *form* from the earliest Greek period to the present. How have these diverse definitions of form influenced philosophical thought?

13. Prepare an essay that explains why the use of the scientific method is the best way to attain knowledge of the external world. Justify your reasons with concrete examples.

REFERENCES CITED

Aldred, C. (1961) The rise of the God-Kings. **In** *The Dawn of Civilization.* Piggott, S. (Ed.) New York: McGraw-Hill Co.

Ayer, A. J. (1959) *Logical Positivism.* New York: The Free Press.

Bailey, N. T. J. (1967) *The Mathematical Approach to Biology and Medicine.* New York: John Wiley.

Barnes, J. (1982) *Aristotle.* Oxford: Oxford University Press.

Bernal, J. D. (1971a) *Science in History. Vol. 1: The Emergence of Science* (3rd Ed.) Cambridge, Mass: MIT Press.

Bernal, J. D. (1971b) *Science in History. Vol. 2: The Scientific and Industrial Revolutions* (3rd Ed.) Cambridge, Mass: MIT Press.

Bonola, R. (1955) *Non-Euclidean Geometry.* (Trans. Carslaw, H. S.). New York: Dover Publications.

Boyer, C. B. (1968) *A History of Mathematics.* Princeton, New Jersey: Princeton University Press.

Clagett, M. (1994) *Greek Science in Antiquity.* New York: Barnes and Noble.

Clarke, D. M. (1982) *Descartes' Philosophy of Science.* Pennsylvania: Pennsylvania State University Press.

Cohen, J. and Steward, I. (1994) *The Collapse of Chaos.* New York: Penguin Books.

Collinson, D. (1987) *Fifty Major Philosophers: A Reference Guide.* New York: Routledge.

Conan, C. A. (1982) *Science: Its history and development the world's cultures.* New York: Facts on File, Inc.

Costantino, M. and Reid, A. (1991) *Leonardo.* Greenwich, Connecticut: Brompton Books, Corp.

Crombie, A. C. (1961) Quantification in medieval physics. **In** *Quantification, A History of the Meaning of Measurement in the Natural and Social Sciences.* Woolf, H. Ed. New York: Bobbs-Merrill Co., Inc.

Crowe, M. J. (1990) *Theories of the World from Antiquity to the Copernican Revolution.* New York: Dover Publications, Inc.

Dilke, O. A. W. (1987) *Mathematics and Measurement.* Berkeley: University of California Press.

Farrington, B. (1961) *Greek Science.* Baltimore, Maryland: Penguin Books.

Grant, E. (1996) *The Foundations of Modern Science in the Middle Ages.* Cambridge: Cambridge University Press.

Hood, M. S. F. (1961) The home of the heroes. **In** *The Dawn of Civilization.* Piggott, S. (Ed.) New York: McGraw-Hill Co.

Hall, A. R. (1981) *From Galileo to Newton.* New York: Dover Publications, Inc.

Hall, M. B. (1994) *The Scientific Renaissance 1450-1630.* New York: Dover Publications, Inc. (originally published in 1962).

Huff, T. E. (1993) *The Rise of Early Modern Science.* New York: Cambridge University Press.

Ifrah. G. The (2000) *Universal History of Numbers.* New York: John Wiley & Sons.

Kierkegaard, S. (1938) *The Journals of Søren Kierkegaard.* (Dru, A. Trans. and Ed.). Oxford: Oxford University Press.

Klein, H. A. (1988) *The Science of Measurement.* New York: Dover Publications, Inc.

Kinder, H. and Hilgemann, W. (1972) *Atlas of World History, Vol. 1* (Menze, E. A. Trans.). New York: Doubleday Anchor Press.

Kline, M. (1959) *Mathematics and the Physical World.* New York: Thomas Y. Crowell Company.

Kline, M. (1972) *Mathematical Thought from Ancient to Modern Times.* Vol. 1 Oxford: Oxford University Press.

Kuhn, T. S. (1957) *The Copernican Revolution.* Boston: Harvard University Press.

Kuhn, T. S. (1970) *The Structure of Scientific Revolutions* (2nd Ed.) Chicago: University of Chicago Press.

Lestrel, P. E. (1997) *Fourier Descriptors and their Applications in Biology.* Cambridge: Cambridge University Press.

Lindberg, D. C. (1992) *The beginnings of Western Science.* Chicago: University of Chicago Press.

Mackay, A. L. (1991) *A Dictionary of Scientific Quotations.* Bristol, U. K.: Institute of Physics Pub.

McLeish, J. (1991) *The Story of Numbers.* New York: Ballantine Books.

Mallowan, M. E. L. (1961) The birth of written history. **In** *The Dawn of Civilization.* Piggott, S. (Ed.) New York: McGraw-Hill Co.

McEvedy, C. (1967) *The Penguin Atlas of Ancient History.* New York: Penguin Books.

Mellaart, J. (1961) The beginnings of village and urban life. **In** *The Dawn of Civilization*. Piggott, S. (Ed.) New York: McGraw-Hill Co.

Neyman, J. (1974) Introduction: Nicholas Copernicus (Mikolaj Kopernik): An intellectual revolutionary. **In** *The Heritage of Copernicus,* Neyman, J. (Ed.). Boston: MIT Press.

Olson, R. (1982) *Science Deified and Science Defied*. Berkeley: University of California Press.

Popper, K. R. (1959) *The Logic of Scientific Discovery*. London: Hutchinson.

Reichenbach, H. (1951) *The Rise of Scientific Philosophy*. Berkeley: University of California Press.

Sarton, G. (1952) *Ancient Science Through the Golden Age of Greece*. New York: Dover Publications, Inc.

Singer, C. (1959) *A History of Scientific Ideas*. Oxford: Oxford University Press.

Sorell, T. (1987) *Descartes*. Oxford: Oxford University Press.

Struik, D. J. (1987) *A Concise History of Mathematics*. New York: Dover Publications, Inc.

Waddington, C. H. (1959) *Tools for Thought*. New York: Basic Books, Inc.

Watson, W. (1961) A cycle of Cathay. **In** *The Dawn of Civilization*. Piggott, S. (Ed.) New York: McGraw-Hill Co.

Wheeler, M. (1961) Ancient India. **In** *The Dawn of Civilization*. Piggott, S. (Ed.) New York: McGraw-Hill Co.

Ziman, J. (1978) *Reliable Knowledge*. Cambridge: Cambridge University Press.

4. TOWARD QUANTIFICATION IN BIOLOGY

To understand a science it is necessary to know its history.

Positive Philosophy
Auguste Comte (1798-1857)

When you can measure what you are speaking about, and express it in numbers, you know something about it, when you cannot express it in numbers, your knowledge is of a meager and unsatisfactory kind; it may be the beginning of knowledge, but you have scarcely, in your thoughts, advanced to the stage of science

Lecture to the Institute of Civil Engineers, 3 May, 1883
Lord Kelvin (1824-1907)

4.1. INTRODUCTION

It is important to realize that morphometrics, in a biological setting, cannot be viewed as detached, or isolated from the conceptual *Weltanschauung,* or worldview, in operation at any particular time. Thus, it will be argued that the much of what consists of the current view of morphometrics continues to be based on outmoded concepts that were shaped in antiquity. Moreover, adherence to this outdated worldview has tended, in part, to hinder the acceptance of newer morphometric approaches in the biological sciences.

In the last chapter, some of the developments of science that led to the birth of modern physics were discussed. The justification for that approach was based on two premises. The first one was that those developments that led to our current views of modern physics have been highly successful and have determined not only the course of physical sciences but science in general. The second reason was that these very successful approaches in the physical sciences have also served as role models for the biological sciences, in spite of considerable difficulties in actual application. Moreover, it will be argued that the, at times, uncritical wholesale acceptance of the methods of physical science may have acted to hinder progress in the biological sciences. This latter point will be examined in Chapter 5 where the development of complexity theory may eventually act to substantially change prevailing views. Nevertheless, as Gerard indicated some time ago:

> Biological theory has been, on the whole, non-quantitative; but only in part is this because of the complexity and fuzziness of the presenting phenomena. It has needed as well ... new mathematical methods to solve the problems (Gerard, 1961:214).

Discussion of these issues will be taken up subsequently in some detail. We now continue our historical survey of science and turn toward biology.

In a sense morphometrics, as well as all the sciences today, are indebted to the Pythagorean School and their "all is number" pronouncement. They clearly had it right, numbers and indeed measurement, remain the essential ingredients of modern science.

Today, the practice of modern biology is also heavily dependent on measurement, although, historically, this represents a rather recent phenomenon. While the roots of modern biology also lie in ancient Greece, particularly with Aristotle and Galen (as discussed in Chapter 3), real progress had to wait for the Renaissance. By 1543, Andreas Vesalius (1514-1564) was able to demonstrate that Galen's anatomy was more animal than human in certain respects, marking the beginning of the decline in Galen's influence. This was followed by William Harvey (1578-1657) who, for the first time, correctly deduced the nature and circulation of blood. Thus, the two aspects: overthrow of the Galenic tradition and rejection of parts of the Aristotelian system, were to play important roles in the emergence of the biological sciences.

4.1.1. Classification of Organisms

The Medieval period led to descriptions of nature from a largely moralistic viewpoint as in the bestiaries and herbals. By the Renaissance, interest in nature was rekindled especially with the discovery of the New World with its foreign fauna and flora. With the increasing success of the development in the physical sciences during the 17th and 18th centuries, there was also a renewed approach toward the scientific understanding of living organisms. Initially, this interest was by amateurs who sought to collect specimens.

However, during 'much of this time, the idea of species as immutable and unchanging was still under the influence of Plato and especially Aristotle. This can be seen in the idea of the *type concept,* that is, the idea of a *type* representing an ideal form against which real specimens were to be compared. Types, in some unspecified way, were presumed to contain all the attributes of a species. From the idea of type, one now had a device, which allowed the construction of a classification of organisms in an Aristotelian fashion. This became a *typology* of organisms, which admitted no change and eventually clashed with the idea of evolution. Today, such classifications are derisively considered as based on *typological thinking.* However, it is important to realize that, prior to Darwin, virtually everyone was a typologist. As a consequence, during the 18th century, the implications of variability, if recognized at all, were not considered important. The presence of variability being relegated to the minor role of 'mere accidents' since they were presumed not to affect the main attributes of form. The idea of evolution, which implied change, did not have fertile soil to grow and prosper. To some extent, remnants of the type concept still appear in the fields of taxonomy, paleontology and archeology.

An example of this was the uncritical application of the Linnean binomial nomenclature to single specimens as they were discovered in the fossil record. Human paleontology still seems to be particularly vulnerable in this regard. By identifying the new discovery with a 'name' (the genus and species tag), implies that it is, somehow, at the mean or center of the population distribution, which may not be the case at all. This is a direct consequence of not recognizing or appreciating the presence of variation in all organisms, living or fossil. One has to always recognize that a single specimen has a probability of being aberrant and not reflecting the average of the population distribution.

Of course, this did not present a problem in 'pre-evolutionary' times,[28] since variation about the mean (the type specimen) was considered essentially zero.

That Darwin was acutely aware of the problem of variation and troubled by it can be seen in his work on barnacles. Darwin found that the variability in barnacles was so great that he had doubts whether two specimens were examples of two species or simply representing variability of a single species (Mayr, 1982).

With Carl Linnaeus (1707-1778), we have the first formalized attempt to systematically classify all living organisms since Aristotle. He took on the considerable task of classifying both the animal and plant kingdoms. He not only defined the groupings called classes and orders, but also further subdivided them into genera and species, forming the *binomial nomenclature* still in use today. For Linnaeus species were constant and unchanging while genera were simply groups of species. Although admittedly, what exactly species were (and are) is still a matter of some controversy. Additionally, there is the issue of what the relationship is between species and genus, beyond being simply useful classificatory devices.[29] Linnaeus and his disciples collected widely to add to his classification scheme. In 1788, the Linnean Society of London was founded (Bernal, 1971). As a testament to the utility of his system, the Linnean classification, although in modified form, continues to be in use today. Nevertheless, the Linnean system was not evolutionary. The idea of the transformation of species was expressed early by naturalists such as Robert Hooke (1635-1703) and John Ray (1627-1705) among others, so the idea of evolution was not a totally unknown one (Singer, 1959). The first suggestions concerning evolution can be attributed to Georges Buffon (1707-1788) although details were necessarily lacking. The ideas of Buffon were examined by Erasmus Darwin (1731-1802) who was the grandfather of Charles Darwin. But it remained for Jean Batiste de Lamarck (1744-1829) to boldly suggest in his book *Philosophie Zoologique,* published in 1809, that species present today were descended from those in previous times. He also indicated that species were modified in the process by closer adaptation to the environment (Bernal, 1971). The idea was startling and unorthodox but won few converts at the time. Moreover, Cuvier vehemently opposed Lamarck's theory of *transformism* as it was then called (Gayrard-Valy, 1994). In retrospect however, it set the stage for what was to come.

Linnaeus had based his classification schema on external parts. Georges Cuvier (1769-1832) on the other hand, used the relationships of the internal structures as a guide to classification. Here Cuvier was rather close to Aristotle with his stress on vegetative and animal function (Singer, 1959). Cuvier's interest in fossils, which were being unearthed with increasing frequency, led him to compare relationships between living and extinct forms. From these he proposed laws of comparative anatomy (Gayrard-Valy, 1994). He, in no small measure, is responsible for what eventually became the field of *paleontology*. Nevertheless, as Linnaeus before him, he believed in

---

[28] Pre-evolutionary times roughly refer to the time before Darwin's seminal work on the origin of species (1859). Although, the notion of evolution was present earlier (Erasmus Darwin, Lamarck, etc), it had little influence on the prevailing *Weltanschauung* until Darwin.

[29] Species are considered as 'real and measurable entities' in contrast to genus and higher categories. Nevertheless, the modern definition of species as a local breeding population, still invokes controversy. At what point is such a population a legitimate species in contrast to being a sub species (or race or variety, to use equivalent terms)?

the constancy of species. The study of internal structures, pioneered by Cuvier, led to the *comparative anatomists* or *morphologists*. This development is taken up again in Section 4.3.

The collecting interest, alluded to earlier, was heightened with the availability of the microscope. Developments such as the microscope opened up a new world that was exploited by Antoni van Leeuwenhoek (1632-1723) and Jan Swammerdam (1637-1680). However, in contrast to the telescope, which played a major factor in astronomy and physics, practical use of the microscope did not become widespread until the work of Koch and Pasteur (see below) 200 years later because of the apparent absence of practical applications. Thus, the lack of application of microscopy, other than for amusement, was one reason for the delay of biology as a scientific discipline (Bernal, 1971). In addition, because of the complexity of biological organisms, processes were bound to be much harder to explain, in contrast to chemistry or physics. We will return to this important point in Chapter 5. The microscope however, did allow for one significant advance, which was the recognition of spermatozoa (in 1686) being responsible for reproduction (finally casting doubt on Aristotle's belief in *spontaneous generation*), although the details initially remained largely unknown.

One of the significant accomplishments in microbiology was the work of Louis Pasteur (1822-1895) who studied the process of fermentation and within it, the role played by bacteria. He discovered that fermentation, decomposition, etc., where all vital processes for which microorganisms were responsible (Singer, 1959). Moreover, the microscope played an increasing role in these discoveries. He found that 'germs' were carried by air currents, which in turn finally provided a mechanism for the transmission of infections. Robert Koch (1843-1910) was the first one to recognize the life cycle of the *anthrax bacillus* providing critically important information on the history of this disease as well providing the necessary impetus that eventually led to *bacteriology* as an independent discipline. Another of Koch's considerable accomplishments, using newly developed staining methods still in use today, was the identification of the *tuberculous bacillus* in 1882 (Singer, 1959).

4.1.2. The Influence of Embryology on Heredity

Another important biological area that was concerned with morphology and had ancient roots was the study of *embryology*. Early interest in embryology, *embryogenesis*, development of sex, etc., can be traced to Aristotle. While Marcello Malpighi (1628-1692) in 1686 was the first to use the chick to illustrate the developing stages of growth, the beginnings of real knowledge of embryogenesis were initiated by the work of Karl Ernst von Baer (1792-1876). Von Baer also used the chick to illustrate vertebral development. He was able to demonstrate the temporary development of gill slits, which eventually led to the famous dictum of: "ontogeny recapitulates phylogeny" attributed to Ernst Haeckel (1834-1919) who was to play a significant part in the acceptance of evolutionary principles in Germany. The development of the fertilized egg (ovum) was demonstrated by Robert Remak (1815-1865) and Johannes Muller (1801-1858) who recognized it as a cell.

From a quantitative viewpoint however, the organization at the cellular and sub cellular levels, had to wait for the proper tools to be developed. While size and number were easily obtained with the increasing use of the light microscope, real quantification required refined techniques not yet available (Gerard, 1961).

Moreover, the actual mechanics involved in embryogenesis, that is differentiation, remained virtually unknown at the end of the 19th century. The prevailing view was that the various organs somehow arose *de-novo* to reside in the ovum. While the process of cell division was recognized earlier, the implications of cell division, *mitosis*, and sexual cell division, *meiosis*, were not really appreciated until the 1880's. Moreover, the relationship of heredity with meiosis was not at all apparent until 1900, although the 1892 research of August Weismann (1834-1914) foreshadowed things to come. He developed the idea of continuity of 'substance', which flowed from parent to offspring. He named this substance the *germ plasm*. The mechanism of how this substance or germ plasm was passed from parent to offspring had already been discovered in 1865 by Gregor Mendel (1822-1884) with his pea plant experiments. Those plant experiments finally demonstrated the correct explanation, *particulate inheritance*, for the transmission of hereditary characteristics from parent to offspring. Mendel's work, (see below) unfortunately, did not become accessible until 1900 (Singer, 1959; Provine, 1971).

Nevertheless, little of these developments, while certainly important in their own right, were to be, in any significant way, quantitative. Although, Mendel's work certainly qualifies as quantitative, if in a rudimentary sense, the majority of biological research up to that time relied primarily on observation and description. The eventual utilization of quantitative methods in biology was to come from two distinctly different directions. This will be taken up in the next section.

4.2. BEGINNINGS OF QUANTIFICATION IN BIOLOGY

The initial application of a quantitative approach to biology can be traced along two general lines. The first emphasis came from the increasing utilization of statistical methods, which started at the beginning of the 19th century. The second one arose from the interest in issues dealing with sex and the transmission of hereditary characteristics and reflected the growing sophistication of these subjects toward the end of the 19th century.

The modern period of biology, and with it morphometrics, may be said to have started approximately 125 years ago. This is the time where mathematical statistics, as an independent discipline from mathematics, finally becomes mature. It is at this time that the foundations of statistics were developed, eventually resulting in such topics as correlation, regression, analysis of variance, etc., methods, which are all routinely used today. One of the interesting aspects of the development of statistics was the implication that it could also provide answers to biological issues. Thus, while there were many individuals who were influential in the development of statistical theory, a number also contributed equally to the resolution of biological issues in the process. As a consequence, the two threads, statistics and heredity, have since the turn of the 20th century became quite closely associated. The following material will illustrate some of the essential inter-connections of the application of the developing statistical methods to the transmission of hereditary characteristics culminating in such fields as quantitative genetics, human population genetics as well as morphometrics.

Biological data, especially from genetics, acted as a catalyst for many statistical developments. It is from this relationship between statistics and biology, initially termed *biometry*, that the seeds of modern morphometrics were sown. Nevertheless, it is not possible to view biometry in isolation. To understand the historical developments that led from biometry to morphometrics, the process has to be viewed within an evolutionary framework that dealt with principles of heredity; in particular, continuous and non-continuous variation. To place this into context, we have to start with the work of Mendel, mentioned earlier, whose seminal 1865 paper correctly outlined the basis for the hereditary transmission of characters from parent to offspring using pea plants.

4.2.1. The Rise of Classical Genetics

Prior to the re-discovery of Mendel's work in 1900, the prevailing view of the transmission of hereditary characteristics was the theory of *blending inheritance* or *pangenesis*. Pangenesis required that the particles containing the hereditary information for every part of the body to be somehow coalesced (blended) in the gonads, and then became incorporated into the reproductive cells. Pangenesis or blending inheritance was also advocated by Darwin as the method of transmission of inherited characters.

Mendel's work, now called *particulate inheritance*, laid the groundwork for the discipline of genetics by providing, for the first time, a testable procedure of how material was transmitted from parent to offspring. He succeeded experimentally were others had previously failed because he proceeded to initially remove some of the sources of variation present in the pea plants prior to carrying out the experiments. For example, by crossing only tall plants and weeding out shorter ones, he was able to produce a relatively *pure line*. This procedure of creating *pure lines* is essential for the second step; namely, the crossing of tall and short plants to be able to successfully demonstrate the 3:1 segregation ratios displaying *dominance*. In 1936 Fisher questioned Mendel's ratios and suggested that they were too exact, that is, too close to expectation, implying experimental bias. See Sturtevant (1965) for a discussion of this matter. Leaving that issue aside, Mendel's work finally provided the correct explanation of how discontinuous variations arose, although, it did not directly address the question of continuous variation. That is, the inheritance of continuous characters, which make up a majority of the heritable material, now called *polygenic inheritance*. Thus, the controversy between continuous and discontinuous variation remained unresolved for the moment.

The problem of variation lies at the heart of Darwinism and the accurate measurement of the form of morphological organisms (to assess this variation) remains a challenge to morphometricians to this day. Thus, in a sense, it was the theory of evolution by natural selection that provided the impetus for statistical and mathematical developments in biology. Although Darwin's predilection, as well as approach, was largely non-quantitative (even though he embraced Malthus' theory of population[30]), the seminal work *On the Origin of the Species* (1859) fostered subsequent studies that were quantitative (Porter, 1986). While Darwin attempted to deal with the causes of variation, the details were not very well understood at the time. Ideas such as differential survival

---

[30] Even if Darwin had read Mendel's paper, it was highly unlikely that he would have recognized its import. This was due, in part, to its mathematical style and Darwin's predilection as a naturalist (Mayr, 1982).

over subsequent generations can only be effectively handled by quantitative methods and these had to wait until the 20th century.

4.2.2. The Roots of Statistical Theory

The origins of statistics can be traced back to the 17th century with work of John Graunt (1620-1674) and William Petty (1623-1687), both of whom were early workers in the field of vital statistics; that is, statistics dealing with political issues like the census, birth rates, deaths, etc. (Sokal and Rohlf, 1969). In fact, the word *statistics* is derived from the word *state*.

The application of statistical tools to biological data, so indispensable today, was initially limited to physics and observational astronomy. In the former, it gave rise to statistical mechanics and in the latter to the theory of propagation of errors. Individuals that stand out in this regard were Pierre Laplace (1749-1827) and Karl Gauss (1777-1855). Gauss made many significant contributions to mathematics, one of which was the development of the *method of Least Squares*, and for recognizing the propagation of errors in observations (Sokal and Rohlf, 1969). However, many years were to pass before the applicability of statistical principles were recognized in the biological and social sciences. Acceptance of statistical methods only arose when the study of biological variation become a legitimate area of concern. Instead of viewing a sample as simply an aggregate of individuals, with the emphasis on the individual, a fundamental change in viewpoint was required. Only when, what Ernst Mayr and others have defined as the 'population approach', became acceptable, could statistical methods be legitimately and profitably applied.

Moreover, also of considerable importance was that the application of statistics revealed the presence of large-scale regularities in nature. This eventually led to the normal or Gaussian probability distribution, the familiar bell-shaped curve. It was first applied in an astronomical context. The variation of the position of stellar objects when repeated observations were involved was interpreted as error due to the imperfections of the instrument as well as due to the senses. This practical problem of how to minimize these observational errors eventually led to the method of Least Squares mentioned earlier (Porter, 1986). The method of Least Squares is now routinely used in linear regression to compute the best fit of a straight line through a set of data points. However, this development did not spur any real interest in the study of variation itself. Interestingly, an early exception here was J. B. J. Fourier (1768-1830), better known for his (1807-1822) researches into the physical transfer of heat (his work will reappear in Chapter 9), who recognized that the mean value by itself conveyed little information and suggested the need to place limits around it. He proposed a standard error of the mean calculation but with much tighter limits than the 5% currently in use (Porter, 1986).

An individual, whose interests in astronomy were to influence developments in biology, was Adolphe Quetelet (1796-1874). He was a Belgian mathematician, astronomer, statistician and sociologist. In 1835, Quetelet demonstrated that human stature was distributed around an average and followed the normal or bell-shaped curve. This was republished in 1869 as *Physique Sociale* and presented his conception of *l'homme moyen* or average man. This notion of the average man was used by Quetelet as a tool for social analysis. It was not only a mathematical convention but also a moral ideal (Porter, 1986). This is not the place to discuss Quetelet's sociology dealing with crime, suicide, marriage, etc., except to note that for Quetelet, ideas dealing with social improvements (enlightenment) should result in a reduction in the extremes about the mean. All the deviations about the mean were flawed and to be viewed as the product of error. For example, the variability in height among Scottish soldiers was due to environmental causes; that is, perturbations due to nutrition, climate, etc. Quetelet was always more interested in the mean values than in causes of variation (Porter, 1986). Ironically, it is the issue dealing with variation (in contrast to the arithmetic mean) that arose from Quetelet's work, which eventually led to major developments in biology such as quantitative and population genetics.

4.2.3. The Controversy over Biological Variation

One individual, who took exception with Quetelet's emphasis on the average or mean, was Francis Galton (1822-1911). He clearly recognized that natural variation was real and not simply error. Galton was a cousin of Charles Darwin whom he greatly admired. In fact, the publication of Darwin's *On the Origin of Species* (1959) had a pronounced effect on Galton's future research endeavors. He was related to the Darwins on his mother's side, and may have felt that he had inherited longevity in life from that side of the family. Reyment suggests that this fact may have influenced his eventual interest in heredity (Reyment, 1996).

In 1869, Galton published his book *Hereditary Genius* in which he attempted to analyze the inheritance of genius. He was certain that genius was an inherited trait and that geniuses displayed superior intellect that separated them out when compared to other humans (Provine, 1971). This eventually led to the movement called *eugenics* (coined by Galton in 1883). In advocating eugenic principles, Galton was, by no means, the first one. Plato in his *Republic* had also advocated such eugenic principles. Galton's idea was that improvement through scientifically directed breeding was possible and desirable. It was predicated on the grounds that physical traits as well as mental abilities were largely inherited. The basis for this was in an early paper: *Hereditary talent and Character*, his first application of statistics to heredity. In it he demonstrated that by tracing the lineages of distinguished men, the number of distinguished offspring was greater than expected compared to the general population (Morgan, 1969). Eugenics for Galton was defined as:

> ... the study of those forces under social control which enhance or impair the inborn qualities of future generations (quoted in Morgan, 1969:94).

Nevertheless, one weakness with the whole eugenic ideal was it's over-emphasis on heredity and its under-emphasis of environmental influences. Both of which are now known to play major roles in modern biology. Initially, eugenics was viewed as unorthodox and did not win any support until the 1890's. While largely discredited today, the movement was of some importance in the earlier part of 20th century and, indirectly, if not directly, responsible for such excesses as Hitler's racial policies. While Galton's eugenic views reflected Victorian upper class ideals, to his credit he never advocated legislation of any kind, or recommended sterilization of the weak and infirm. He merely pointed out possible consequences of breeding as he saw them (Porter, 1986).

Galton was also responsible for the terminology 'nature-nurture'; That is, both environmental as well as genetic aspects making up the *phenotype*. Nevertheless, he held the view that inherited properties were much more important, especially in the achievement of individuals. This ran counter to the prevailing Victorian notion that achievement was attainable by hard work. His studies of human twins suggested to him that even complex abilities were largely determined by heredity.

However, the obstacle for Galton (as for others) was how these hereditary characteristics were transmitted to the next generation. This was before the re-discovery of Mendel's work in 1900. The only theory available was blending inheritance or *pangenesis*. Darwin (in the absence of knowledge of Mendelian genetics) also had to fall back on pangenesis to account for hereditary differences. In 1869 with the publication of *Hereditary Genius*, Galton already questioned pangenesis and with it the Darwinian notion of gradual evolution by natural selection. Galton suggested that evolution was discontinuous and based on a series of steps or *sports* or discontinuous leaps (saltations) which were in themselves stable. Galton attempted to test pangenesis experimentally by transfusing the blood of different varieties of rabbits on the assumption that the gemmules circulated through the blood stream. If the offspring of the rabbits were somehow a hybrid type then Darwin's theory would be proved. The experiments turned out to be negative and led Galton to conclude that pangenesis was, in fact, incorrect. Galton wrote in the *Proceedings of the Royal Society* that blood was not a fundamental source of heredity (Morgan, 1969). Darwin replied, writing in *Nature* that the experiments did not really deal with pangenesis. This set the stage for the ensuing controversy between continuous and discontinuous variation. However, the issue could not be really resolved at that time since there really was no clear evidence that gemmules were, in fact, present in the blood (Provine, 1971). Moreover, the Galton idea of sports was also initially equivocal although it does anticipate the concept of mutations as they are now called, but which were also not clearly identified as such at the time.

By 1875, Galton began to search for a more satisfactory theory of heredity that would bolster his conviction of the importance of discontinuous variation. He had observed that the mean value of characters such as stature remained roughly the same from generation to generation. However, the offspring of the tallest had, on the average, shorter offspring than the mid-parent value[31], while the shortest individuals had, on the average, taller offspring than the mid-parent value. This phenomenon has been called *regressing* toward the population mean (Mayr, 1982). His break with Darwin on the issue of variation was complete by his 1889 book *Natural Inheritance*.

According to Provine there were two primary reasons: [1] that sports (saltations) were stable and therefore unchanging and [2] belief in the principle of regression toward the mean.[32] Galton based his arguments on the following:

---

[31] Mid-parent value refers to the average of the parents.
[32] Galton is credited with the first formulation of what we would today term linear regression. It was initially called the 'law of ancestral heredity' by Galton (see Porter, 1986).

> In a population whose general characters remain constant over a period of generations, each character nevertheless exhibits some variability each generation. Yet, the range of this variability does not change from generation to generation. Thus, the exceptional members of the population cannot produce even more exceptional offspring, on the average, or else the range of variability of the character in question would expand markedly (Provine, 1971:19).

and on:

> ... his first statement of the law of regression: then mean ... of progeny was displaced from the general mean in an amount proportional to the displacement of their parents. The mean displacement of the offspring, however, *was always less* than that of the parents... (Porter, 1986:286-287; italics mine)

Based on these assumptions, Galton implied that exceptional members could only appear then via the so-called saltations or discontinuous leaps. As Provine put it:

> Looking at the effects of blending inheritance and regression, Galton decided that sports must be the only source of evolutionary variation. Darwin, looking at blending inheritance, decided just the opposite—that sports could play no role in evolution (Provine, 1971:23).

This controversy, between continuous and discontinuous variation, was heightened with the rediscovery of Mendel's work in 1900. Moreover, it was to have a major influence on the development of population genetics. This controversy was not to be effectively settled until the evolutionary synthesis in the 1930's.

Galton had a way of collecting statistics on any number of subjects to support or refute theories. Material for such data collections came from the first ever, anthropometric laboratory set up by Galton at the International Health Exhibition in London in 1884 (Morgan, 1969). The collection of data for statistical analysis has now become a major part of biology. He contributed to meteorology as well as recommending the use of fingerprinting for forensic purposes. Among the many accomplishments of Galton, three stand out. The first one was his emphasis on variability. His second one was his invention of correlation, which Karl Pearson later improved upon. The third was the principle of regression, alluded to earlier, for which he was also responsible. All are now statistical concepts of central importance in the biological sciences. He used the method of regression to estimate the height of an individual from knowing the heights of his relatives. Galton's *Natural Inheritance,* while in some respects suspect and today dated, dealt with regression and correlation and was to have a major impact on future biometricians such as Karl Pearson and W. F. R. Weldon and has been considered by some to represent the starting point of biometry and, in a sense, also of morphometrics.

In his will, Galton had provided funds for a research chair at the University College, London. A position later occupied by Karl Pearson (1857-1936). Pearson a brilliant mathematician, perfected the coefficient of correlation and developed the chi-square goodness-of-fit test. Not the least of his many contributions was the idea of using a sample drawn from the population. Pearson shared Galton's outlook regarding eugenics as well as minimizing the effects of environment on the phenotype, especially with regard to mental qualities. However, on the issue of evolution, in contrast to Galton, he was a follower of Darwin's continuous evolution as early as 1883 (Provine, 1971). Also in the camp of the biometricians was W. F. R. Weldon (1860-1906). In Weldon's view, Galton's sports could be effective in exceptional cases, but the small variations arising from natural selection were sufficient to explain evolution (Provine, 1971). By 1901-1902, the evolutionary controversy between the Mendelians (favoring noncontinuous variation) and the Biometricians (arguing for continuous variation) was in earnest. This controversy, because of the strong personalities involved, was to hold back the evolutionary synthesis for some three decades. In 1901, criticisms by William Bateson (1861-1926), an avowed Mendelian, of one of Pearson's papers in the Proceedings of the Royal Society eventually led to, among other things, a new journal. The new statistical journal *Biometrika* was set up in 1901 by Pearson and Weldon, with the assistance of Galton, a journal still active today (Porter, 1986; Provine, 1971). Besides the new journal, differences between the Mendelians, typified by Bateson, and the biometricians, initially Weldon and later Pearson, led to public clashes. Facts that played a major part in the affair were the personal antagonisms between the participants and the relative complexity of the newly proposed statistical arguments. Moreover, Bateson was never very competent in mathematics and continuous variation required a statistical bent.

Hugo de Vries (1848-1935), one of the re-discoverers of Mendel's work in 1900, is also credited with the discovery of mutations, published by him in 1901. Bateson and de Vries had become friends (although later this friendship cooled) as they shared a belief in noncontinuous evolution, as well as a mutual dislike for the biometricians (Provine, 1971). Thus, with de Vries mutation work, the Mendelians presumed that their position had been greatly strengthened vis-à-vis Darwin's gradual evolution by selection. It was even suggested that all speciation was a result of mutations alone. To quote Mayr:

> The leading textbook of genetics during the de Vries era (Lock, 1906:144) summarizes the thinking of the Mendelians in the statement: "Species arise by mutation, a sudden step in which either a single character or a whole set of characters together become changed." (Mayr, 1982:547)

One of the fallouts of initial mutation theory of de Vries, which was favorably perceived by many biologists as an alternative theory to natural selection, was the decline of Darwinism. The concept of gradual evolutionary change by natural selection operating on continuous characters was for a time eclipsed. This was, in part, because of the lack of clear evidence for natural selection and the fact that many viewed Mendelism and noncontinuous variation as synonymous. In an attempt to bolster the biometry position, experiments utilizing mice were initiated by A. D. Darbishire, a student of Weldon. Unfortunately, these experiments provided, much to the discomfort of Weldon and Pearson, more evidence of the validity of Mendelian segregation ratios. These experiments were also subsequently validated by William Castle (1867-1962) in the U. S. Almost the only notable exception to the controversy was the mathematician G. U. Yule (1871-1951), who in 1902 suggested, using mathematical arguments, that perhaps Mendelism and Darwinian evolution might yet be compatible. He predicated his argument on the possibility that Mendel's factors might show variation, that is, be present in small steps (what would now be termed *multifactorial inheritance*). In other words, the variation present could be due to a number of factors. Nevertheless, this valuable contribution to the literature did little to ameliorate the acrimony and confusion between the two sides (Provine, 1971).

4.2.4. The Evolutionary Synthesis

It took another sixteen years before Yule's suggestions were finally to bear fruit in the hands of Fisher. Sir Ronald Fisher (1890-1962) continued the development of modern statistical theory initiated by Pearson. He also made extensive contributions to the design of experiments developing analysis of variance models as well as contributing techniques such as discriminant functions to what is today called multivariate statistics. Many of his major contributions were to the field of genetics. In 1918, Fisher published *The Correlation Between Relatives on the Supposition of Mendelian Inheritance*, what was to become a classic paper. This paper was the first one to demonstrate that the results of the biometricians could be explained in terms of Mendelian inheritance (Provine, 1971). Today it is well known that Mendelian particulate inheritance represents only a small portion of what is naturally occurring variation. Continuous variation as exemplified by *e.g.*, stature, as well as a myriad of other characters, represents the majority of biological, as well as behavioral traits such as intelligence to name only one of the physiological traits to receive considerable if, at times, controversial attention. One notable characteristic of these continuous traits is that they generally follow a Gaussian distribution.

Fisher approached the problem by defining a measure of variability—the variance—of a measurable character, *e.g.,* human stature. He then broke down the total variance into its constituent parts such that the total variance, or phenotypic variance, (V_P), is equal to a genetic component (V_G) and an environmental one (V_E):

$$V_P = V_E + V_G.$$

[4-1]

This partition formulates the famous problem of *heredity* versus *environment*. Note that this approach deals with the *variance* of a character, not with its actual metrical measurement. The actual situation is considerably more complicated as it is known that the environment can affect the genetics of a character. Thus, it is more correct to re-write equation [4-1] as:

$$V_P = V_E + V_G + 2\,\mathrm{cov}_{GE},$$

[4-2]

where the $2\,\mathrm{cov}_{GE}$ term is the covariance between the genetic and environmental effects.[33]

As Fisher suggested, other influences can obscure the actual genetic or phenotypic similarity between relatives in addition to environmental effects. The latter having been demonstrated with artificial selection experiments. These genetic effects were

---

[33] Covariance is a measure of the degree by which two variables vary together. It is related to correlation which is the covariance divided by the product of the standard deviations of the two variables. Thus, it can be viewed as a measure of the correlation between heredity and environment.

dominance (as discovered by Mendel and displayed with the familiar 3:1 ratio) and the presence of genetic interaction or *epistasis*. He considered these two effects to be *non-additive*. Thus, in addition to the environmental variation (V_E), the total phenotypic variation of a trait was now composed of two major terms, an additive factor (V_A), and a non-additive one. The non-additive one being broken down further into two terms, dominance (V_D) and epistasis (V_I) or interaction. That is:

$$V_P = V_E + V_A + V_D + V_I .$$ [4-3]

Detailed discussions of these variance components can be found in Falconer (1960), Mather and Jinks (1971), Cavalli-Sforza and Bodmer (1971), as well as many other books on population genetics. From these and other considerations, Fisher concluded that:

> The simplest hypothesis ... is that such features as stature are determined by a large number of Mendelian factors (Quote in Provine, 1971:145).

While the contribution of each of these factors was individually small, the total effect satisfactorily explained the distribution of measurable characters. Thus, there was no need to postulate the presence of major environmental effects, which led Fisher to conclude that:

> ... it was unlikely that so much as 5% of the total variance is due to causes not heritable (Quote in Provine, 1971:147).

This last statement still invokes some controversy today with respect to the percentages of environmental versus genetic influences on a trait. Although Fisher's paper was difficult to read because of the mathematical arguments, it set the stage for the eventual accommodation between the Mendelians and the Darwinians. By the 1930's, the remaining puzzles had started to fall into place. In 1930, Fisher published *The Genetical Theory of Natural Selection* in which he attempted to produce a complete theory of evolution. In it, he mathematically showed that Darwin's pangenesis (blending inheritance) was fatally flawed in that it cut the heritable variation in half each generation. For Darwin's theory of natural selection to work required the maintenance of variation each generation. Darwinian evolution required new variation each generation and today it is readily apparent that the process of *meiosis* provided the essential mechanism for this maintenance of variation. Fisher showed that this variation, in fact was, preserved with Mendelian inheritance. Fisher's work paved the way for the eventual synthesis between Mendelism, Darwinism and Biometry.

The second major figure in this synthesis was Sewall Wright (1889-1988). Wright's views were influenced by the experimental researches of William Castle (1867-1962) with whom he worked from 1912-1915. Here Wright conducted experiments on the inheritance of coat color in guinea pigs. He found that multiple interacting factors were responsible for coat color. He eventually focused on the effects of inbreeding in Mendelian populations. Wright approached the problem of the distribution of traits (as gene frequencies) and evolution from the viewpoint of inbreeding and his newly developed method of path coefficients (Provine, 1971). This was in marked contrast to Fisher who used a statistical approach based on a variance components model as discussed earlier. Thus, Wright in a paper entitled *Evolution in Mendelian populations* (1931), while indicating his general agreement with Fisher's arguments, took exception on one point, which was that he felt that small rather than large populations, were central with respect to evolutionary change. He felt that in small populations random effects (accidents of sampling) could appreciably alter the outcome of gene frequencies through time (Wright, 1970 and elsewhere). This effect was eventually termed *random genetic drift* and is now considered one of the forces of evolution in conjunction with mutation and selection. Random genetic drift developed as an outcome of Wright's inbreeding studies.

J. B. S. Haldane (1892-1964) was the third pivotal figure responsible for the evolutionary synthesis. His book *The Causes of Evolution* (1932) became a classic because of its examination of Darwinism from a 20th century perspective of the knowledge of heredity and variation. To which he added his considerable mathematical analyses. In a quote that stressed both quantification and elucidation of process, concepts that have equal relevance for morphometrics today as they had then for the incipient field of population genetics, Haldane stated:

> A satisfactory theory of natural selection must be quantitative. In order to establish the view that natural selection is capable of accounting for the known facts of evolution, we must show not only that it can cause a species to change, but that it can cause it to change at a rate which will account for present and transmutation (Quoted from Wright, 1968:3).

Thus, the work of R. A. Fisher, Sewall Wright and J. B. S. Haldane can considered as primary (there were, of course, many others who contributed along the way) for the development of the mathematical background to evolutionary theory using, in a large part, statistical principles. This was made possible, in large part, by the recognition that one can view the genetics of biological organisms as composed of organizational units called genes that operate at the molecular level (the genotype), and which eventually produce the populational level (the phenotype). Their work led to the establishment of the current field of population genetics and to what became the so-called *evolutionary synthesis*. This synthesis arose because of the increasing recognition that evolution was multifaceted and could not be fully explained by a single aspect like selection or the presence of sports, *i.e.,* mutations. By the 1930's and 1940's evolutionary theory consisted of four processes which were: [1] mutation, the fundamental source of variation, [2] natural selection, [3] migration and [4] random genetic drift. These four factors, while not necessarily of equal importance or perhaps the only ones, nevertheless, continue to remain the keystones of evolutionary theory today.

Genetics, and in particular population genetics, have made good use of quantification with the result that evolutionary processes are now relatively well understood. Thus, population genetics is one of the most developed or 'mature' disciplines in biology rivaling the physical sciences in some respects (Smith, 1968). In addition to the application of statistical methods in biology leading ultimately to genetics, other areas of biology have also benefited from the increasing application of mathematical approaches; that is, approaches other than statistics, in an attempt to elucidate biological processes.

Moreover, numerous problems arise in biology that can be most usefully treated quantitatively. The methods have been available from the physical sciences for a good long time now. Consider the forces involved in mammalian locomotion. These involve tension and compression of bones viewed statically (when the organism is at rest) as well as dynamically. Alternatively, one might want to compute the energy expended during locomotion. There are a myriad of other factors as Maynard Smith has cogently pointed out, which include the rate of heat loss of muscles, the rate at which oxygen is supplied to the tissues, constraints arising from the size of bones and muscles, etc. (Smith, 1968). Each of these aspects lends itself, at least on one level, (in a rather straightforward way one might add) to evaluation with quantitative methods. That is, it is not difficult to describe some of these individual processes such as blood flow, etc., although computations may become rather sophisticated requiring differential equations. One early worker was Alfred Lotka (1880-1949) whose book, initially published in 1920's, laid out a program that eventually led to what is now called *theoretical biology* (Lotka, 1924). Another early worker was D'Arcy Thompson (1860-1948) whose work will be examined in some detail in the next section.

However, on a deeper level, all these and numerous other factors alluded above operate simultaneously during locomotion rapidly making a 'total solution' very complex. It is this state of affairs that has limited many areas of biology and continues to given them an essentially descriptive character. Some of the reasons for this have already been alluded to such as the complexity of biological organisms, which precludes easy answers except in the simplest cases.

One way to attempt to describe these aspects in totality is to view biological processes as a *system*. System analysis and the building of models are areas of increasing interest to biologists and will be taken up in the next chapter where we will specifically address the issue of viewing the morphological form as a system. Now we can turn to one of the central tenets of this volume: The quantification of the biological form.

4.3. THE QUANTITATIVE STUDY OF FORM

The above two sections, represent an attempt to historically trace the intertwining threads that led to quantification in biology. This section focuses on concept of *form* and how the increasing numerical characterization of morphological attributes led to the development of modern morphometrics. For the last century, or more, there has been little questioning of the validity of the measurement systems used to describe the biological form. In fact, for early biologists the very act of taking measurements was viewed as an unattractive and unnecessary endeavor.

Clearly, times have changed, and the use of quantitative methods is now widespread. Moreover, with the rapidly increasing utilization of powerful computers (available even now at the PC level), the ability to analyze as well as reduce large data sets has become routine and extensive developments since the 1960's have occurred in an attempt to simplify these large data sets. These developments have led to an increasing application of statistical methods to biological data sets. The success of these methods is now undoubted. However, even those who routinely apply these multivariate procedures rarely question the appropriateness of the measurements being utilized or necessarily always understand the appropriateness of the statistical tests being applied to these data. Thus, a number of basic questions can be raised about the utility of the measurement system being used to derive these data sets. A concomitant query that needs to be asked is whether some results to date need re-evaluation in the light of this issue. However, first we need to explore the history of measurement of form.

4.3.1. Early Developments in the Study of Morphology

Aristotle can be credited with being the first real morphologist. He clearly saw that organisms could be grouped together according to a *unity of plan* based on resemblance of external and internal characteristics (Mayr, 1982). The term *morphology* was introduced by Johann Wolfgang von Goethe (1749-1832) about 1807 (Mayr, 1982). Emphasis on external characters also continued, leading to the study of *systematics* or *taxonomy,* the latter word being introduced by Augustin Candolle (1778-1841) in 1813 (Singer, 1959).

Richard Owen (1804-1892), a disciple of Cuvier, extended Cuvier's work in comparative anatomy and paleontology.[34] He independently developed a system of classification, which, however, did not become generally accepted. He introduced the word *dinosaur* at the 1841 meeting of the *British Association of Science* (Gayrard-Valy, 1994). One of his claims to fame was his opposition to Darwinism (Singer, 1959). This emphasis on the morphology of both the external characters as well as the internal structure, in conjunction with eventual quantification, can be considered to have played a major role in what was to become biometry and still later morphometrics.

A Dutch anatomist, Peter Camper (1722-1789) was probably one of the earliest individuals to try to quantitatively describe aspects of anatomy. His book published posthumously in 1792, *Dissertation on the Natural Varieties which Characterize the Human Physiognomy* was to have a marked influence on measurement in the biological sciences; and particularly, on physical anthropology (Greene, 1959). He attempted to define numerical differences in skull shape between ethnic groups using degrees of *prognathism* (Hays, 1972). Camper developed two metrics. One was the *facial line* and the other the *facial angle.* The facial line was drawn from the forehead to the upper lip. The facial angle is formed by the intersection of the facial line with a line drawn from the *auditory meatus* to the base of the nose. These two measures anticipate the eventual widespread use of the conventional metrical approach (CMA) some 100 years later (Chapter 8). A serious problem, not understood at the time, with such measures is that they were generally applied as measures to individual skulls or heads. The collection of samples (to assess the presence of variability) rather than individuals was rarely contemplated. Moreover, adults were illustrated together with juveniles, thereby largely ignoring changes occurring due to growth.

The common use of indices or ratios can be traced to the work of Magnus Retzius (1842-1919), who introduced the length-breadth, or cephalic index (cranial index on skulls). This was intended to provide an estimate of skull shape in *Norma verticalis (i.e.,* looking down on the top of the skull) of living individuals (Hjortsjö and Lindegard, 1953).

---

[34] In 1838 the word paleontology was coined by the geologist Charles Lyell (Singer, 1959)

The use of such indices can be viewed as a forerunner of *multivariate* statistics in that it represents an attempt to view morphological shape differences with more than one variable. The extension from two to three variables can be seen with the use of the cranial module defined as length plus breadth plus height divided by three (Martin, 1928). The drawbacks of such measures as indices will be examined in more detail in Chapter 8.

The use of linear measures, angles and indices steadily increased during the 19th century, culminating in the Frankfurther Verstädigung (Frankfort Agreement) of 1884. The Frankfort Agreement of 1884 was further codified with the Monaco Agreement of 1906, which attempted to lay the groundwork for a series of standardized measurements, many of which remain central in the numerical characterization of the craniofacial complex. This agreement formally adopted such metrics as useful classificatory measures. Karl Pearson's group (Section 4.2.3) subsequently used these measures in a series of papers starting with the founding of the journal *Biometrika* in 1901 and roughly ending in the mid 1930's. These papers mark the beginnings of multivariate morphometrics (Chapter 7). This preoccupation with measurement also led to a German guidebook for craniofacial measurement, the well-known *Lehrbuch der Anthropologie*, used with skull materials (Martin, 1928).

Pearson was also responsible for the Coefficient of Racial Likeness (CRL). The CRL was intended to summarize with a single measure complex forms (Pearson, 1926). In a sense, the CRL anticipated multivariate statistical methods. The intent of the CRL was to utilize multiple measures (based on means) to determine whether two groups were, in fact, significantly different from each other.[35] The CRL also stimulated Fisher, leading him to the development of discriminant functions (Section 7.2.1). Finally, Penrose (1954)[36] attempted a re-formulation of Pearson's CRL by partitioning it into size and shape components. These two components, size and shape, play a role in boundary morphometrics (Chapter 9).

4.3.2. From Morphology to Morphometrics

Also spanning much of this period of increasing quantification in biology was a player of major importance for the modern development of morphometrics. The influence of D'Arcy Thompson (1860-1948) is still being felt today. In his classic work, *On Growth and Form* (originally published in 1917) Thompson re-interpreted the growth and structure of organisms in mathematical and physical terms. This was a substantial advance in quantifying biology but had little influence at the time the book was published. A major reason for the reluctance in accepting quantitative approaches was that, by then, biology was primarily viewed in terms of comparative anatomy coupled with principles drawn from evolutionary theory, a largely descriptive endeavor

---

[35] Objections raised against the CRL included the lack of correction for: [1] significant differences in variance of the measurements chosen and [2] significant correlation between variables. Although, it should be mentioned that Pearson was aware of the issue of correlation (Fisher, 1936).

[36] Unfortunately, Penrose's re-formulation did resolve the issue of correlation. This had to wait for the work of Mahalanobis (Section 7.2.2).

rather removed from Thompson's views.[37] For Thompson, changes in form over time (growth) were primarily due to the actions of physical forces; and hence, reducible to the type of force diagrams characteristic of Newtonian physics. Thompson's view of form was couched in physical terms:

> The form, then, of any portion of matter, whether it be living or dead, and the changes in form which are apparent in its movements and its growth, may in all cases alike be described as due to action of force (Thompson, 1942:16).

Nevertheless, this rather mechanistic approach to growth has turned out to be not only much more difficult to apply in actual practice, but does not take into account the phylogenetic history that led to the biological form.

One of Thompson's many accomplishments was the notion that organismal change over time affected the whole organism and not just the constituent parts. This led him to the method of *coordinate transformations*, which while seemingly based on mathematical principles, were not effectively carried out by him. It remained for others such as Medawar and Bookstein to explore these possibilities (Chapter 8). In an early monograph dealing with the measurement of growth and form, Zuckerman questioned:

> ... whether precise numerical formulations of temporal size and shape can be derived from some fundamental biological law of growth; or whether numerical generalities are purely empirical measures which, in D'Arcy Thompson's words, in turn become steps in an analytical process (Zuckerman, 1950:435).

It is perhaps more useful to turn the question posed by Zuckerman around and ask *not* whether "numerical formulations of temporal size and shape can be derived from some fundamental biological law", but rather whether biological processes *can* even be derived, in some sense, from an analysis of size and shape. It precisely this question that remains a challenge for morphometrics. Thus, to investigate this further we need to start with the fundamental notion of form. This is taken up in the next two chapters.

KEY POINTS OF THE CHAPTER

This chapter summarized some of the threads that led to the beginnings of quantification in biology. The idea of variation of organisms, so central to modern biological thought, was scarcely considered in pre-Darwinian times. Until Darwin in 1859, the Aristotelean concept of type held supreme. The majority of all research in biology, whether pre- or post-Darwin, remained overwhelmingly descriptive. It was developments in statistics in the middle of 19th century that eventually stimulated the increasing use of quantitative methods in biology. The earliest discipline to profit from the application of statistical theory was the field of heredity. Mendel's work on pea plants eventually led to modern genetics. The threads, in terms of measurement, can be traced back to the work of Camper at the end of the 1700's, although, little progress was

---

[37] Nevertheless, Thompson was solely a functional morphologist who rejected Darwinism (natural selection) in favor of a view that claimed that organisms (because of their inherent plasticity) could readily adapt to immediate functional constraints, making an analysis in evolutionary terms untenable, or at the very least unnecessary (Mayr, 1982). This rather one-sided view is now largely discredited.

made until the late 19th and early 20th centuries. The use of multiple measurements to describe the morphology of skulls led to an increasing quantitative approach in biology, leading to such fields as numerical taxonomy as well as being an impetus to the separately developing field of multivariate statistics.

CHECK YOUR UNDERSTANDING

1. Discuss the concept of speciation and examine the role that quantification plays in it.

2. What role did studies of heredity play in the adoption of a quantitative approach to biology?

3. Trace the development of the microscope and examine how it influenced the use of numerical methods in biology.

4. How did the study of morphology by anatomists at the end of the 19th century impinge on developments in statistics?

5. How did the work of Fisher, Haldane and Wright affect the development of evolutionary theory?

6. Discuss the concept of variation in the light of the work of the Mendelians and the biometricians.

7. What is meant by the evolutionary synthesis? How did quantitative methods play role in it?

8. Trace the idea of the type concept from the time of Aristotle. Does it still have an influence on modern biological thought?

9. Contrast the basis of typological thinking with thinking in evolutionary terms.

10. Define the following terms: *Blending inheritance, particulate inheritance* and *polygenic inheritance.*

11. How does the concept of variation affect the outcomes of blending inheritance and particulate inheritance? What can you infer from these observations about the process of evolution (natural selection)?

REFERENCES CITED

Bernal, J. D. (1971) *Science in History. Vol. 2: The Scientific and Industrial Revolutions* (3rd Ed.) Cambridge, Mass: MIT Press.

Cavalli-Sforza, L. L. and Bodmer, W. F. (1971) *The Genetics of Human Populations.* San Francisco: W. H. Freeman and Co.

Fisher, R. A. (1936) The Coefficient of Racial Likeness and the future of craniometry. *J. Roy. Soc. Anthrop. Inst. G. Brit. Ire.* **66**:57-63.

Falconer, D. S. (1960) *Introduction to Quantitative Genetics.* New York: The Ronald Press Co.

Gayrard-Valy, Y. (1994) *Fossils—Evidence of Vanished Worlds.* New York: Harry N. Abrams, Inc. Pubs.

Gerard, R. W. (1961) Quantification in biology. **In** *Quantification, A History of the Meaning of Measurement in the Natural and Social Sciences.* Woolf, H. (Ed.) New York: Bobbs-Merrill Co., Inc.

Greene, J. C. *The Death of Adam.* (1959) Ames, Iowa: Iowa State University Press.

Hays, H. R. (1972) *Birds, Beasts and Men.* Baltimore: Penguin Books, Inc.

Hjortsjö, C. and Lindegard, B. (1953) Critical aspects in on the use of indices in physical anthropology. *Kungl. Fysiograf. Sälls Förhandl.* (Lund). **23**:1-9.

Lotka, A. J. (1956) *Elements of Mathematical Biology.* New York: Dover Publications, Inc.

Martin, R. (1928) *Lehrbuch der Anthropologie.* Stuttgart: G. Fischer. (Re-issued in 1957).

Mather, K. and Jinks, J. L. (1971) *Biometrical Genetics.* (2nd. Ed.) London: Chapman and Hall.

Mayr, E. (1982) *The Growth of Biological Thought.* Cambridge: Harvard University Press.

Morgan, R. W. (1969) Sir Francis Galton (1822-1910). **In** *Some Nineteenth Century British Scientists.* Harr, R. Ed. London: Pergamon Press.

Pearson, K. (1926) On the Coefficient of Racial Likeness. *Biometrika.* **16**:328-363.

Penrose, L. S. (1954) Distance, size and shape. *Ann. Eugen.* **18**:337-343.

Porter, T. M. (1986) *The Rise of Statistical Thinking 1820-1900.* New Jersey: Princeton University Press.

Provine, W. B. (1971) *The Origins of Theoretical Population Genetics.* Chicago: University of Chicago Press.

Reyment, R. (1996) An idiosyncratic history of early morphometrics. **In** *Advances in Morphometrics* (Marcus, L. F., Corti, M., Loy, A., Naylor, G. J. P. and Slice, D. E. (Eds.) New York: Plenum Press.

Singer, C. A (1959) *History of Scientific Ideas.* Oxford: Oxford University Press.

Smith, J. M. (1968) *Mathematical Ideas in Biology.* Cambridge: Cambridge University Press.

Sokal, R. R. and Rohlf, F. J. (1969) *Biometry.* San Francisco: W. H. Freeman and Co.

Sturtevant, A. H. (1965) *A History of Genetics.* New York: Harper and Row, Pub.

Thompson, D.W. (1917) *On Growth and form.* Cambridge: Cambridge University Press.

Thompson, D.W. (1942) *On Growth and form. (2nd Ed.)* Cambridge: Cambridge University Press.

Wright, S. (1931) Evolution in Mendelian populations. *Genet.* **16**:97-159.

Wright, S. (1968) Contributions to genetics. **In** *Haldane and Modern Biology.* Dronamraju, K. R. (Ed.). Baltimore: The John Hopkins Press.

Wright, S. (1970) Random drift and the shifting balance theory. **In** *Mathematical Topics in Population Genetics,* Kojima, K. (Ed.).

Zuckerman, S. (1950) The pattern of change in size and shape. *Proc. Roy. Soc. (Lond.)* B. **137**:433-442.

5. COMPLEXITY, SYSTEMS AND MODELS

Essentia non sunt multiplicanda praeter necessitatem.
Entities are not to be multiplied beyond necessity.

Attributed to William of Ockham (*ca.* 1285-1349)

A l'heure actuelle la Biologie n'est qu'un immense cimitière de faits, vaguement synthesés par un petit nombre de formules creusés.
At present biology is nothing more than an immense graveyard of facts, vaguely held together by a few empty formulas.

Rene Thom (1923-)
In E. C. Zeeman (Ed.) Catastrophe Theory, 1977

5.1. INTRODUCTION

In Chapter 4, the historical development of quantification in biology was briefly reviewed leading to the study of morphology and eventually to the beginnings of the numerical analysis of form. Initially however, form was solely characterized in descriptive terms. The roots of the quantitative approach can be discerned in the work of Camper in the 1700's and found full expression in the work of D'Arcy Thompson. In this chapter, we begin to take up the quantitative representations of form in some detail and in the next chapter; an admittedly heuristic model of form is presented.

Advances in the biological sciences often proceed on a number of fronts, which include [1] development of theoretical frameworks (formal model building), [2] development of appropriate tools and [3] applications based on various techniques (using data) to solve problems. All three of these approaches are conceptually linked and need to proceed simultaneously if progress is to be made. Generally, specific data-oriented problems tend to spearhead the need for new techniques leading to new algorithms. These in turn, may lead to a re-evaluation of accepted theory or assist in the development of new formal models. As Wilks has succinctly stated:

> Finally, I would like to comment on what we may regard as perhaps the highest form of quantification in science. This consists of the mathematical models, which describe essential features of the quantitative relationships inherent in vast systems of measurements. A model of this type not only provides a relatively simple and elegant scheme for describing a system of measurements, which have already been made, but also serves as a dependable instrument for predicting the outcome of further measurements which would belong to the system if they were made. Examples of such models are Kepler's laws of planetary motion, Newton's more general laws of motion... (Wilks, 1961:12).

One of the undoubted successes of the scientific method has been the ability to explain as well as predict the pattern of physical behavior, as seen in the development of modern physics.[38] Consequently, some have suggested that physics is unique and should be relegated to a special worldview, in contrast to other scientific disciplines. It is suggested here that this is a specious argument and does not hold upon closer scrutiny. There is no logical reason to suggest that physics, and the methods used within it, require special pleading. Physics, as a discipline is not in any way unique, it has simply been initially easier to 'model' phenomena here than in other sciences, specifically the biological or social ones. Modeling in physics has been successful because the laws of physics were presumed to be relatively simple, one of the simplest being $F = ma$; thus, while physics is 'simple', other fields studying nature, were considered more complex (Bak, 1996). However, even in physics things are not as seemingly straightforward as they once were. Consider Newtonian physics, Poincaré was able to show in 1890 that Newton's laws of motion did *not* provide a general solution when more than two bodies are involved. That is, in the case of the sun, moon and the earth, what became the *three-body problem,* there is no analytic solution, only approximation methods. Here Poincaré anticipated what is now called chaos theory (Cambel, 1993; Briggs and Peat, 1989). Moreover, recent developments in physics, vis-à-vis string theory, black holes, etc., suggest that even hallowed concepts such as the Einsteinian space-time continuum may require considerable re-assessment in the future.

Nevertheless, the failure to successfully apply the modeling approach of physics to biology, is in large measure due to the complexity of biological phenomena. Complexity itself is a complicated issue and will be taken up in the next section. It is the lack of models in the biological sciences that, by necessity, have reduced these disciplines to being largely descriptive. One needs to remember that all new sciences start with description. It is only subsequently, as a science begins to mature, that the goals shift toward explanation in contrast to simply description. Nevertheless, description continues to provide the raw data used for the development and verification of those models aimed at explanation.

Because of the extreme complexity of biological systems, those simple models that *are* available tend to be under-specified as well as display glaring oversimplifications (Bailey, 1967). Such models have tended to be of limited use. The difficulties inherent in attempting to formulate models that can account, even in a limited way, for the diverse morphological structures that have arisen from conception to the adult organism, are immense. Rene Thom in discussing his catastrophic model of morphogenesis, indicated that:

> In biology, if we make exceptions of the theory of population and formal genetics, the use of mathematics is confined to *modeling* a few local situations (transmission of nerve impulses, blood flow in the arteries, etc.) of slight theoretical interest and of limited practical value (Thom, 1983:114; italics added).

Moreover, this is a situation that has not materially changed much in spite of the fact that:

> ... the problem of the integration of local mechanisms into a global structure is the central problem of biology... (Thom, 1983:154).

---

[38] For example, the ability to make reasonably long-term predictions of planetary orbits, (given that no anomalous events occur) has facilitated space travel and shows the undoubted success of physics.

This 'global structure' needs to be viewed as a model of biologic process. In a sense, Thom's work can be considered one of the early precursors of complexity theory. Related, but applicable in a wider context, to the model of growth (whether viewed in deterministic, stochastic or catastrophic terms); is the need for an adequate numerical characterization of form which becomes, in one way or another, the raw data for model building. Those few examples of models in the biological sciences that are available (Heinmets, 1969), either tend to be of very low predictive power or extremely restricted making them very specialized and, thus, of limited interest as alluded to by Thom above. If theory building in the biological sciences is to advance, then emphasis needs to shift from description to process. It is in this shift that systems and the modeling of systems, becomes of central importance. First, we need to examine some of the reasons why explanations of biological processes have remained, with few exceptions, comparatively limited and non-quantitative in contrast to other fields, such as physics, and this leads to the study of complexity.

5.2. COMPLEXITY

Consider the following rather diverse examples dealing with process: [1] what will the weather be in Los Angeles on February 19, 2020, [2] when will the stock market Dow Jones averages reach 20,000, [3] when will the next magnitude 7+ earthquake hit Los Angeles, and [4] what will the facial profile of one's daughter look like at 22 years of age given that she just turned three, etc. What these, and any number of other examples that could have been chosen, share, is that none of them admit easy (if at all) answers and all are exceedingly complex in character. Moreover, models aside, there are no analytic [39] procedures available that can provide precise answers to the above examples. While other approaches (*e.g.*, stochastic[40] ones or simulations[41]) have been occasionally applied, their utilization remains limited at the moment. With respect to the last example above, dealing with biological growth or *morphogenesis*, there is also no guarantee that such predictions will be accurate (Section 5.2.3).

Clearly, it is a given that life forms, even single celled ones are, at one level, quite complex. However, to define something as complex is, in itself, not terribly useful or informative. Accordingly, what is needed is a more formal definition of what complexity is and, more specifically, a framework that allows biological processes to be more profitably viewed.

> The ability of such systems to resist analysis by traditional reductionist tools of science has given rise to what is now called the sciences of complexity, involving the search for new theoretical frameworks and methodological tools for understanding these *complex systems* (Casti, 1997:34; ital. in the original).

---

[39] Analytic refers to the use of algebraic instead of geometric methods to describe numerical relationships between variables. In addition, functions with power series expansions are analytic, *e.g.*, a Fourier series (Chapter 9).

[40] Stochastic refers to the use of random variables, *i.e.*, instead of analytic variables, which are point estimates with no variability; stochastic variables are based on a probability distribution with generally finite variance (*e.g.*, see Levin, 1969).

[41] Simulation refers to the construction of a particular mathematical model (as a simplification) to estimate the characteristics associated with *e.g.*, a biological process such as species competition or organismal growth.

Complexity theory allows for the exploration of similarities in such diverse areas as biological systems, ecosystems, economic systems, behavioral and learning systems, etc. Although, complexity is displayed in so many ways, and in so many different fields of endeavor, that definition becomes difficult. Nevertheless, some of the aspects of complexity can be provisionally identified in a heuristic fashion (Cambel, 1993). The following aspects, others could also have been chosen, are aimed at biological systems but not solely confined to them:

---

1. Complexity is present in all systems, whether natural or man-made.
2. Complexity arises whether the biological system is large or small.
3. The larger the number of components, the more likely complexity arises.
4. Complexity occurs in both energy-conserving and dissipating systems.
5. The different parts of complex systems are always interrelated.
6. The complex system contains positive or negative feedback loops.
7. Complex biological systems are open with respect to the environment.
8. Biological systems obtain nutrients, energy and information externally.
9. Complex biological systems tend to evolve in an irreversible manner.
10. Complex biological systems are dynamic and not in equilibrium.

Table 5.1. Some aspects of complexity.

Another way to view complexity is with hierarchical levels of organization. For example, one can organize or arrange a biology system as composed of: organelles, cells, tissue, organ, organism, population, and finally, as an ecological community. How these spatially organized units are related is better known for some of the links than for others. Focusing on the organism, except for the obvious simple characteristics, there is still much that is unclear about what the common fundamental organizing principles are that determine the form or *bauplan* of all animals (Thom, 1983).

To further illustrate the nature of complexity, it is perhaps helpful to think in terms of two *levels* for the lack of a better term. Level 1 is 'horizontally layered', so to speak, based on *scale* and Level 2 is 'vertically layered' based on *rank*. Additionally, the horizontal layers can be viewed as orthogonal, or independent, to the vertical layers. Thus, for an organism within the Level 1, any one of the horizontal layers can be viewed as composed of subunits that are roughly of the same *scale*. That is, within Level 1, one can consider an organism as composed of bones, muscles and tissue such as lungs and heart, etc. Level 2 is *hierarchical*, that is it consists of different vertical layers. These vertical layers comprising Level 2, can be viewed as starting with DNA developing into cells, which develop into organs and eventually into whole organisms, etc. This has been defined by Bartlett as proceeding:

... from the molecular biology level to the organization and functioning of a single cell, from there to the structure and functioning of parts of an organism, such as the muscles or the nervous system, thence to the structure and behavior of the entire organism, and, finally to the statistical properties of whole populations (Bartlett, 1968:208).

While much research has focused on aspects within the horizontal layers (Level 1), quite successfully one might add, the integration of the horizontal levels with the vertical ones (Level 2) has been less than desirable, this being a consequence of complexity. Thus, the analysis of complex biological systems continues to be a major challenge for biologists.

5.2.1. Complex Adaptive Systems

While complexity can be defined in a number of different ways, Holland defines complexity as having a number of attributes, which together form *complex adaptive systems*, or *cas* (Holland, 1995). Note that the emphasis is on adaptation.[42] This leads to the question of what exactly is *cas*?

A simple definition of *cas* is the *composition of interacting elements, which can be described by a set of rules*. According to Holland, such systems can be distinguished by the presence of particular properties like: [1] aggregation, [2] non-linearity, [3] flows and [4] diversity. Others could probably also be enumerated. Each of these four ideas will be briefly outlined.

The property of aggregation has two meanings for Holland (1995). The first one refers to grouping of similar objects. This definition has already been encountered (Section 1.2). A way to simplify complex systems is to organize them according to similarities in patterning. Objects such as dogs, trees, etc., can be placed into categories. These categories can also be termed equivalence classes. Of considerable importance is that this categorization process—the aggregate—is selective. What this means is that those details that are irrelevant for the questions being addressed can be ignored. For example, in the process of building an equivalence class of dogs, the fact that they all contain four limbs becomes irrelevant in terms of providing information that distinguishes different dog varieties since they all contain that characteristic. This attribute of selectivity becomes particularly important in the construction of models, of which aggregation is one of the main components (models will be explored in greater detail below).

The second meaning is more related to *cas*, *s*pecifically, the complex aggregation of less complex agents. Consider an ant nest. Behavior of the individual ant is largely genetic and stereotyped, slight changes in the environment surrounding the ant can readily cause its death. Yet, the aggregate, the ant nest (composed of individual ants), is

---

[42] In complexity theory, adaptation has a considerably broadened definition that applies to many fields, economics, business, etc., not just biology. However, biological adaptation is more narrowly defined in terms of fitness. That is, adaptation, or fitness, is a function of natural selection. The simplest linear model is W=1-s, which states that as the selection coefficient (s) increases in value, the adaptation or fitness decreases. The average fitness of a population can then be, theoretically, set up as a Hardy-Weinberg equilibrium: $W=p^2W_{11}+2pqW_{12}+q^2W_{22}$ (Li, 1955; Lestrel, 1967). Nevertheless, it is important to realize that this is a greatly simplified approach as real populations rarely see equilibrium.

highly adaptive and can survive in the face of numerous hazards. In other words, the ant nest, like a city has a life of its own. Now consider another example, the lower jaw. It is composed of subunits such as the dentition and muscles. These subunits, in turn, are composed of smaller units such as cells. The interrelationship of all structures at different levels (scaling) produces a mandible that is identified with a unique human individual. Each of these units (the mandible) and subunits (a molar) can aggregate, in turn, in their own right and be composed of other aggregates such as certain cell types, and potentially, down to the genome. This is the implication of complex aggregation of less complex agents (agents are defined subsequently).

The second property delineated by Holland is non-linearity. Most of the models derived from mathematics are linear in nature. These include the fitting of straight lines to data (linear regression) as well as being the basis of many multivariate statistical procedures (for example, general linear models or GLM). Considerable success has been attained with these approaches, in spite of the fact that most biological phenomena are often non-linear in behavior. Unfortunately, once one comes to grips with the complexity of phenomena, nonlinearities become the rule and the application of *cas* requires a non-linear approach (see Holland, 1995:15-23 for examples).

The third property of flows refers to much more than fluid flow. One can speak of the capital flows in economics or traffic flow, etc. In a biological context, one might think of the flow of messages along neurons, or mRNA. In an embryological context, biological growth from the beginning of the zygote takes place because of such message flows, even if some of the details remain unclear. It is apparent that growth, for example, exhibits largely orderly behavior, which may be emulated with a series of *rules*. Consider simple stimulus-response relationships, these may be described with an '*if...then*' format; equivalently, it may be possible to track how information flows in biological systems such as growth. We will return to this issue subsequently.

The last property, diversity, is, perhaps, more straightforward. Consider the diversity of insects in a rain forest. However there is one caveat here, it must be emphasized that there is no *a-priori* reason for adaptation to necessarily always lead to diversity and with it, an increase in complexity. While increases in complexity abound in many lineages, the lack of it is also evident *e.g.,* sharks and crocodiles have remained virtually unchanged over millions of years. Here adaptation led to an evolutionary pathway that was, and has remained optimal. According to Holland, the persistence of organismal diversity in biology is partially due to the filling of niches (vacated due to extinction) by organisms that are already partially adapted. In Holland's words:

> ... the system typically responds with a cascade of adaptations ... This process is akin to the phenomenon of *convergence* in biology. The ichthyosaur of the ancient Triassic seas filled much the same niche as the porpoise in modern seas (Holland, 1995:27).

Moreover, the diversity of *cas* in biology is not only dynamic (*i.e.,* evolving) but persistent in the sense that when disturbances arises, such as species extinction, other species arise to 'repair' the system and generally assure continued persistence. To quote Holland once more:

The diversity observed in *cas* is the product of progressive adaptations. Each new adaptation opens the possibility for further interactions and new niches. ...What mechanisms enable *cas* to generate and maintain temporal patterns...? (Holland, 1995:29).

While answers to Holland's question can be hypothesized, they are difficult to demonstrate experimentally. Nevertheless, modeling, specifically simulation, is one approach that may provide possible directions for future research, if not specific answers. We will return to the building of models subsequently. Another technical term introduced by Holland is that of *agent*.[43] Agents are the basic building blocks of complex systems. These are the cells, species, etc., which are:

... constantly organizing and reorganizing themselves into larger structures ... At each level new emergent structures would form ... (Waldrop, 1992:88)

Agents at one level act as building blocks for agents at a higher level. Consider a group of nucleic acids, lipids, proteins, etc., being the constituents making up a cell. Accordingly, *cas* is based on these agents. These agents are modified or adapt, resulting in organizational re-arrangements over time in response to evolutionary processes. Such complex systems rarely reach equilibrium, adaptation generally provides for only short-term changes and not long-term equilibrium. If this 'equilibrium' is not a necessary consequence of evolution, does this not also suggest the need for other mechanisms to complement the four classical forces of evolution: mutation, natural selection, migration and drift? Possible additional mechanisms are listed in Table 5.2 (modified from Mainzer, 1997).

---

1. Ability to self-reproduce
2. Maintenance of metabolism
3. Capability of optimal adaptation
4. Presence of feedback loops

Table 5.2. Mechanisms of complexity.

Self-reproduction is essential for continuing existence at the individual and species levels. Metabolism is necessary to counter the increase in entropy predicted by the second law of thermodynamics. Adaptation leads to increased complexity, by aiming toward optimal conditions even if they are never reached in an equilibrium sense. Finally, the presence of feedback is a crucial component. Such feedback loops can be seen in the switching mechanism of genes where certain genes turn on and off other genes; or in predator-prey relationships. Feedback mechanisms allow for response to changing environmental conditions, thereby increasing organismal survival. Thus, it has

---

[43] Agents refer to the active elements that comprise all systems. These could be antibodies in an immune system, cells in a biological system, communities in an ecosystem, automobiles within a transport system, or even companies in a large city. The agents are active in the sense that they interact in complex ways (Holland, 1995).

become increasingly apparent that the presence of *cas* suggests that adaptation in terms of the forces of natural selection, mutation and other evolutionary mechanisms, can only partially account for complexity. Other factors such as emergence and self-organization must also be taken into account.

5.2.2. Properties of Emergence and Self-Organization

Emergence is the property by which the interaction of elements leads to newly organized matter. The following simple example should make this clearer:

> Weather is an emergent property: take your water vapor out over the Gulf of Mexico and let it interact with sunlight and wind and it can organize into an emergent structure known as a hurricane (Waldrop, 1992:82).

The process of morphogenesis can also be viewed as an emergent property, albeit a particularly complex one; one that is composed of a hierarchy or many levels of complexity as described earlier. That is, in very cursory way: the DNA code directs the building of amino acids, which eventually are responsible for protein structure. These proteins then determine configurations called cells, which, in turn, aggregate into tissues leading to other emergent structures such as organs, and so on. At each level of complexity, new properties appear. The organizing principles that control emergence at each level may also fundamentally change in character. A challenging element of emergence is its, at times, unpredictability. That is, particular aspects of emergence (often time-dependent) display discontinuity. This has also been termed a phase transition, which is in a word, *chaotic*.[44] An example is the sudden appearance of a tornado, or a landslide, or in morphogenesis, the moment when tissue differentiation occurs. Thus, complexity then starts with the description of emergence. What remains, and can be considered one of the major issues in biology, is the determination of the laws that control emergence. Thus, it seems increasingly apparent that one of the most challenging issues facing the biological sciences today has to do with the development of models, which provide some measure of explanation of *how* such biological systems persist (see Holland, 1998).

Ever since Ludwig Boltzmann (1844-1906) attempted to relate the biological evolution to principles derived from chemistry and the laws of thermodynamics, an intractable issue has arisen. If the second law of thermodynamics predicts decay and disorder with a consequent increase in entropy, how was it possible for Darwinian evolution to generate an orderly development of living systems with increasing complexity (Mainzer, 1997)? The presence and persistence of complexity suggests that:

> Boltzmann's thermodynamics as well as Newton's mechanics are insufficient to model the emergence of complex order, and thus the origin and growth of living systems (Mainzer, 1997:90).

A related issue is how this increase in complexity is even possible in the face of models of equilibrium. With respect to this latter point, consider that much of the

---

[44] Defined as unpredictable behavior arising in a system which extremely sensitive to variations in initial conditions. Some examples are turbulent flow, long-range weather patterns, and cardiac arrhythmias.

literature of classical population genetics, vis-à-vis the work of Fisher, Haldane and Wright (briefly reviewed in Section 4.2.1.), is predicated on reaching some kind of evolutionary equilibrium. This notion of equilibrium models as being somehow the norm is not limited to biology. Such models also exist in geology and economics. Moreover, the widespread use of equilibrium models is readily understandable. They can be more easily constructed (especially linear ones) in contrast to models that exhibit complexity (non-linearity). Non-linear models were initially considered largely intractable. Another factor that played a part was that such systems, especially biological systems, were presumed to be in balance (Bak, 1996). That is, in terms of evolutionary time, it is not surprising that with the span of human years, systems tended to look as either very slow changing or constant with time. However, such equilibrium models eventfully become static and cannot readily incorporate change from an evolutionary point of view. This can again be seen as the problem of order out of randomness. Nevertheless, it has to be admitted that no theoretical framework is yet available that handles such complex non-equilibrium based systems.

The implication that evolutionary paths will lead necessarily to equilibrium via the Hardy-Weinberg equation was also questioned by Holland. Particularly vexing, was how order (emergence) can arise given the independence of genes advocated by Fisher, not even considering seemingly random events such as mutations. As Waldrop put it:

> A single gene for green eyes isn't worth very much unless it is backed up by the dozens or hundreds of genes that specify the structure of the eye itself (Waldrop, 1992:165).

This led Holland to consider that perhaps genes formed such groups or clusters. This necessarily leads to the question of how do such clusters, as building blocks, ultimately determine the adult phenotype. Holland gives an example of the building blocks that determine the unique human face. Consider a face composed of ten components (eyes, eyebrows, nose, lips, etc.). Now consider each feature composed of ten alternatives (blue, green, hazel, etc., for eye color). This leads to 10x10 or 10^2 building blocks. Continuing, Holland indicated:

> Because there are 10 alternatives ... we can construct any of $10^{10} = 10$ billion distinct faces from these 100 building blocks. Almost any face we encounter can be closely described by an appropriate choice from the set of 100 building blocks (Holland, 1995:36-37)

This suggests that a relatively few elements (although the actual number may be substantial) are needed to account for the variability seen in biological organisms. In terms of morphogenesis, perhaps such a series of clusters of cells is responsible for differentiation. As Waldrop put it:

> If a cluster was coherent enough and stable enough, that it could serve as a building block for some larger cluster, cells make tissues, tissues make organs, organs make organisms, organisms make ecosystems ... (Waldrop, 1992:169).

In this way, emergence or *cas* would arise. This led Holland to modeling or specifically, to simulation of a genetic algorithm. In brief, the model required three things. It had to allow for sexual reproduction (meiosis), crossing over and for the presence of occasional mutations.

> Reproduction and crossover provided the mechanism for building blocks of genes to emerge and evolve together ... (Waldrop, 1992:174).

Holland was able to demonstrate that genes, indeed, did form clusters and if these clusters had above average fitness, their percentages would increase in a population. However, the genetic model was not sufficient to explain *cas*. It needed something else, a feedback system; that is, learning if you will. After all, as Darwin had recognized, it is the environment that is the ultimate arbitrator on how species adapt and survive or become extinct. Holland added a second approach consisting of if-then rules, which were mediated with a feedback system. The incorporation of rules and the feedback system led to a model that began to hint at how, starting with essentially randomness (chaos), one can evolve structure (order). This approach represents some of the incipient work, which may eventually lead to a paradigm shift (to use Kuhn's phase) in the biological sciences. The details can be found in Holland (1995) and elsewhere, and the interested student is well advised to pursue these ideas.

5.2.3. Morphogenesis and Complexity

A number of fundamental problems are attendant with morphogenesis and some of these can be profitably viewed in terms of complexity (Section 5.2.1). At a basic level, biological morphogenesis implies the formation of new structures, as well as the modification and change of existing morphologies with time (see Section for 5.4.5 for an approach to modeling growth). With respect to the first point, as noted by Sheldrake:

> The first problem is precisely that form comes into being ... new structures appear which cannot be explained in terms of an unfolding or growth of structures which are already present in the egg at the beginning of development (Sheldrake, 1995:19)

A second issue is that developing systems can *regulate*. That is, during early development, damage to the system can be compensated for with the consequence that more or less normal growth is established. Removal of one of the cells at the two-cell embryo stage gave rise to a small but intact sea-urchin (Sheldrake, 1995). However, as development continues, this capacity is eventually lost as the differentiation of tissues and organs is completed. Nevertheless, this regulating capacity implies the presence of feedback loops. A third issue is *regeneration*, since many organisms have the capability to replace damaged structures, *e.g.*, the lens from the iris margin of a newt. In lesser capacity, the ability to heal wounds is also an example of regeneration. A fourth issue is act of *reproduction*, specifically, *parthenogenesis*, where a detached part of the adult becomes a new organism, a result which can be considered as a "part becoming a whole" (Sheldrake, 1995:21). These issues have led to a variety of explanations such as vital factors (recall Bergson, Section 3.5.2), morphogenetic fields (Waddington, 1957) and genetic programming. However appealing as the latter two notions may be, in a similar way to blending inheritance (Section 4.2.1) earlier, they lack sufficient explaining power in terms of mathematical treatment, at least for now. As Sheldrake has indicated:

> It implies that the fertilized egg contains a pre-formed programme which somehow specifies the organism's morphogenetic goals and coordinates and controls its development toward them. But the genetic programme must involve something more than the chemical structure of DNA, because identical copies of DNA are past on to all cells; if all cells were programmed identically, they could not develop differently (Sheldrake, 1995:21).

Progress has been made toward answering some of these very challenging issues dealing with morphogenesis and these are taken up in Section 5.4.5. We now turn to other approaches that might be potentially useful for resolving some of these issues.

5.3. SYSTEMS THEORY

Historically, the systems approach can be traced to Newton's *Principia* (1687). In his famous work, Newton viewed the solar system as composed of the sun, the planets, the moon and comets with interactions between them due to gravitational forces. His *mathematical representation* of this 'system' was based on his three laws of motion and the 'law' of universal gravitation. In schematic form, this is depicted in Figure 5.1.

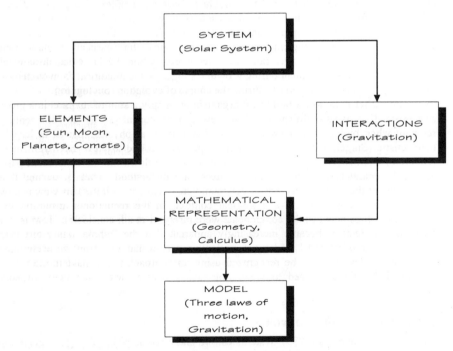

Figure 5.1. The Newtonian system.

What exactly then is a system? Moreover, what are the properties that determine a system? Finally, how do you go about doing such an analysis with biological systems?

Systems analysis is the study of the structure and behavior of sets of interacting elements. Another similar definition of a system is:

> ... a set of objects together with relationships between the objects and their attributes (Weinberg, 1975:63).

The distinguishing characteristic of systems analysis is the treatment of the system as a whole rather than composed of pieces, which are separately analyzed with little thought of their interrelationships with other pieces. One way to study systems is by building models. Note that that the Newtonian explanation of the solar system is such a model. This leads to the question of what is a model. One definition of a model is that of a mathematical representation of the system (models and modeling will be examined in more detail below).

5.3.1. Systems Applications in Biology

Barriers to the application of system analysis to biological systems include the complexity of the subject matter, lack of testable theories about process and the, until recent, lack of methods that could be brought to bear on complex biological systems. Taking a cue from *cas*, we now know that we need to view the biological form not only as a system, but also as a complex adapting system. In line with the central focus of this volume; namely, the biological form, such a biological 'form-system' displays both complexity and adaptation over time. In other words, while the individual biological form adapts to the environment during growth (ontogeny), the biological form-system— the aggregate—adapts and persists during the course of evolution (phylogeny).

The earlier notion of a biological organism as simply a sum of its parts is a gross and misleading oversimplification. That view can be considered as an 18th century remnant when biology was viewed, in a similar fashion as physics, that is, as largely mechanistic (Singer, 1959). That traditional approach tended to view the organism in *reductionist* terms. That is, the organism can be broken down into its constituent parts and each separately analyzed. Once the pieces are understood, it was presumed that knowledge of the whole entity was largely complete (Clements, 1989). This view is now known to be inapplicable in a biological context. In the reductionist program, the interrelationships between the components were rarely, if at all, considered. That is, for biological systems it became increasingly apparent that the "whole was more than simply the sum of its parts". Moreover, critical properties that arise from these complex interrelationships can only be recognized using an approach that considers the whole system. Thus, there is need to view biological processes within a systems theory framework.

5.3.2. The Systems Approach

According to Clements (1989), systems can be classified as physical (*e.g.*, engineering structures), abstractions (*e.g.*, stock market transactions), human based (*e.g.*, any one of numerous human activities), and natural (*e.g.*, biological organisms and systems). Other systems, no doubt, arise, but for purposes here, only the last one is of interest from a system theory viewpoint. The approach of systems analysis requires a number of steps, which are listed in Table 5.3. Problem formulation generally involves a deep understanding of the biological system and the environment in which it is embedded. The system needs to be subdivided into smaller elements, the components of which need to be identified. The behavior and interactions of the components also needs to be assessed. A mathematical model can then be constructed to yield insight into the workings of the system.

---

1. Identify and formulate problem
2. Identify components involved in the system
3. Identify the interactions in the system
4. Developed the mathematical model
5. Relate the model to the system
6. Draw conclusions about the system from the model
7. Test against data and refine the model

Table 5.3. The systems approach.

One particularly fruitful way of approaching biological processes is to view them in *systems* terms. System analysis and the building of models offers considerable potential; however, systems analysis is still rarely utilized in the biological sciences. In the next sections, we take up the issue of viewing the aspects of the morphological form as a system to be modeled.

5.4. THE DEVELOPMENT OF MODELS

In previous sections, a number of reasons were presented for justifying the need for models. One of these reasons was predicated on the need to model the biological process as a system. However, biological systems are distinctly different from systems developed in the physical sciences, in that they represent *cas* (Section 5.1.1). Because of this complexity mentioned earlier, procedures have been developed based on systems analysis, leading to models. The idea of a model is to discern the underlying processes characteristic of biological systems, composed of numerous interacting elements. For example, a biological system can be envisaged as composed of sub-structures that are dependent on each other (consider the lower jaw and its interconnections, consisting of muscles, etc., with the cranial base). One is interested in not only how such a system works but also how it adapts to changes imposed on it (often over time). Ultimately, a good (effective) model will allow for prediction, in a generalized way, of those system changes. While mathematical models have been remarkably effective in the physical sciences, they have been less successful in the biological sciences, undoubtedly due to, as stated earlier, the complexity of living organisms.

A model has been defined as consisting of two elements: [1] the choice of variables to be included and [2] the relationship between these variables (Saaty and Alexander, 1982). This approach, however, is in need of some amplification (Section 5.4.1). While there seem to be as many diverse definitions of models, the one preferred here, views modeling as a procedural approach. That is, one needs to view *modeling* as a process and the *model* as the result or end product. Modeling has a long history in the physical sciences and especially engineering, but remains largely under-utilized in the biological sciences.

5.4.1. Model Building

Model building is defined here as the process of describing natural systems in mathematical terms. Two questions that one may legitimately raise now are: [1] what exactly is a model and [2] how are such models constructed? An answer to the first question, couched in heuristic terms, is as follows. A formal definition of a model is: A set of equations that are a *representation* of a system such that the solutions provide an adequate *understanding* of aspects of the systems' behavior.

In brief, we start with a biological system from which we gather sufficient data to formulate a model. Simplification is generally indicated; otherwise, it may become impossible to build the model.[45] Next, the model is analyzed and conclusions are drawn. These conclusions form the basis for interpretations about the biological process under consideration. Finally, these conclusions about the biological process are tested or verified against new observations and new data. This leads to changes, and if necessary, refinement and improvement of the model.

That is, enough data must be available but too much data is generally not practical. Clearly, this is directly related to the question of how many variables are needed in a model. There is yet no ready answer to that question. Nevertheless, while experimentation and many computer runs may be needed, the ability to build good models is definitely a skill that will be increasingly needed to elucidate biological processes.

The word *model* is derived from 'modus' (a measure), which implies a change of scale (Aris, 1978). A particularly good example of a model in this sense is a map. A cartographic map is a model that represents actual topography, but in a simplified way. Details not considered relevant are missing (also actual physical space on a map mitigates against displaying much detail). Thus, scale influences detail. This issue of scale will re-appear again in the discussion on dynamic modeling (Chapter 6).

As long ago as 400 BC Plato recognized that words could be used as models. In a sense, the noun 'tree' is a model or 'picture' of an actual tree. While not terribly precise, it does communicate something that is recognizable by others (Cross and Moscardini, 1985). A, perhaps, better example is the Japanese Kanji character, *ki*, (borrowed from China) for tree, because it really looks like a tree! It is not an exact replica but it contains the essential elements (trunk and branches). In an analogous fashion, Newton's laws can also be viewed as 'idealized' models of physical reality upon which Einstein improved. Thus, the development of models is an essential aspect of science in general. Put somewhat differently, model building is an essential step in the formulation of theories (Aris, 1978). Nevertheless, it must be emphasized that all models are a representation of reality, not reality itself. The reluctance to reject the eventually discredited Ptolemaic theory of the solar system in favor of the Copernican system is an example of the error of confusing the model with reality. In this particular case, reasons for this reluctance to

---

[45] Nevertheless, while simplification is desired, the opposite may also be problematic; *i.e.,* building a model with an insufficient number of variables, *e.g.,* the use of CMA to characterize form (Chapter 8).

accept the correct model having more to do with religious preconceptions than with scientific observation.

5.4.2. Types of Models

The simplest model is probably a *flowchart*. This is a graphical display using boxes to represent individual components, the subsystems making up a total system, with subsystem interactions being described with arrows. While useful in a preliminary fashion to outline the basic elements of a model, flowcharts are largely qualitative in character. A more appropriate and useful model is a mathematical one that describes the subsystem interactions with equations, thereby making it quantitative. Such models play significant roles in computer science and engineering as well as a host of other disciplines.

Mathematical models represent both idealization and simplification. This is an inevitable consequence. The test of a model is always how close it mirrors reality. This process of model building is illustrated with a block diagram in Figure 5.2 (this model will be examined in some detail in Section 5.4.3). Of crucial importance here is that the process displayed in Figure 5.2 is an iterative one. That is, one starts with an idealized simple model, which initially does not provide a good correlation with reality, by gradually refining the model (adding variables, equations, etc.) one eventually should be able to model the biological process of interest. In one sense then, the modeling approach simply represents another approach of the scientific method (Clements, 1989). In judging the validity of a model, Aris (1978) envisioned the modeling process to advance roughly along the following lines given that some theoretical knowledge was available to guide and refine the approach (Table 5.4). However, in terms of actual application this can become a complex and sophisticated venture.

1. Preliminary idea(s)
2. Initial model
3. Design of experiment(s)
4. Testing
5. Evaluation
6. Interpretation
7. Revised ideas
8. Leading to subsequent models.

Table 5.4. The modeling process.

It should be noted that as Aris indicated:

> The revision of ideas and development of models is not necessarily in the direction of greater complexity or an increasing number of parameters. Progress may be toward simplification and the reduction of the number of adjustable constants ... Nevertheless, part of the judgment of a model will lie in whether its constants can be found from independent sources and combined to give a convincing picture in the interactive situation (Aris, 1978:23).

There are numerous types of models in the literature, as well as various definitions as to what models are, making matters difficult. The varied uses of the word model have engendered considerable confusion (Aris, 1978). Moreover, the approach, or technique, to be used depends on the type of model under consideration, thus, mathematical models may be: [1] largely *descriptive* (a majority in biology are) and [2] explanatory or *conceptual* (Kac, 1969). The Copernican view of the solar system would be a conceptual model, in part because of its predictive power. Most models probably lie between these two extremes in that they contain elements of both. Models may also be *static* (*e.g.,* in equilibrium) or *dynamic*; that is, changing over time. Models may be *analytic* (deterministic) or *stochastic* (include random components). Moreover, models may be described as *discrete* (using difference equations for solutions) or *continuous* (using differential equations).

According to Meyer (1984), three steps are involved in mathematical modeling. These are: [1] *formulation*, which focuses on the issues of interest, the identification of factors which are relevant and how to describe them in mathematical terms; [2] *manipulation*, this refers to the procedure of setting up the relationships in mathematical form and solving them;. and [3] *Evaluation*, or testing the validity of the model. How closely does it accord with reality? Models by their very nature are never exact and should be viewed as approximations.

Two approaches to model building can be discerned. Approach I is based on knowledge of the underlying biological process to develop the mathematical equations and build the model. Such models are theory-based and tend to be more complex in character. In contrast, approach II is largely descriptive in nature; that is, theory-less, so to speak. That is, one uses the observed pattern in nature and fits it with a function. Such models are generally descriptive and have limited and specific application. That is, they are not comprehensive or global in nature. They provide a very limited picture of reality but can serve as a starting point for more sophisticated approaches. Accordingly, with respect to approach II, one such biological model involved the use of differential equations to solve ecological issues such as predator-prey relationships (Levin, 1969). Another area is the development of functions to describe growth.

5.4.3. Modeling: A Simple Example

Consider some of the exponential models that have been applied to growth, specifically human growth. Interest in modeling human growth goes back to Thomas Malthus (1766-1834) who in 1798 suggested that geometric (exponential) growth of populations would eventually outstrip food production, the rate of which was assumed to be arithmetic, that is, linear. We can build a simple model if we ignore complexities such as the effects of the *demographic transition*. The demographic transition refers to the reduction of death rates followed by the dramatic fall in birth rates, a now well-documented phenomenon, which has led to population stability in the industrial nations. This however, is still not the case for the developing or third world countries. The basic idea is that the rate of change of a parameter (which may be negative such as radioactive decay, or positive such as human growth) is directly proportional to the instantaneous value of the parameter (Dym and Ivey, 1980). That is, the rate of change of N (size of a collection of objects) with t (time) is directly proportional to the instantaneous value of N:

$$\frac{dN(t)}{dt} = \lambda N(t), \qquad [5\text{-}1]$$

where λ is the proportionality constant. Ignoring the details here since they are available in textbooks, it can be readily shown that the solution to the first order differential equation [5-1] is:

$$N(t) = N_0 e^{\lambda t}, \qquad [5\text{-}2]$$

where e=2.71828, the base of natural logs (ln) and N_0 is the initial population size. If we plot the ln of $N(t)$ instead of $N(t)$ against time we can create a semi-log plot which is a straight line with the slope of λ. That is:

$$\ln N(t) = \ln N_0 + \lambda t. \qquad [5\text{-}3]$$

If we now were to modify equation [5-1] to include a stabilizing factor to limit population growth, this would yield:

$$\frac{dN(t)}{dt} = \lambda_1 N - \lambda_2 N^2 \qquad \lambda_1, \lambda_2 > 0, \qquad [5\text{-}4]$$

where λ_1 is the potential or uninhibited growth rate and the ratio λ_1/λ_2 is the maximally obtainable population size at which the population would reach stability. Letting $dN(t)/dt$ vanish, it follows that:

$$\lambda_1 N - \lambda_2 N^2 = 0, \qquad [5\text{-}5]$$

or rewriting as:

$$N_m = N_{max} = \frac{\lambda_1}{\lambda_2}. \qquad [5\text{-}6]$$

Re-writing Equation [5-4] in terms of N_{max} and λ_1, and thereby eliminating λ_2 yields:

$$\frac{dN}{dt} = \lambda_1 N \left(1 - \frac{N}{N_m} \right).$$ [5-7]

Equation [5-7] is the exponential model modified to include a factor that is the proportional difference between the maximum population and the instantaneous population (Dym and Ivey, 1980). Finally, solving equation [5-7], subject to the initial condition of $N(t=0) = N_0$, we arrive at:

$$N_t = \frac{N_m}{1 + \left(\dfrac{N_m}{N_0} - 1 \right) e^{-\lambda t}}$$ [5-8]

For $t=0$, equation [5-8] yields the initial condition, while for $t \rightarrow \infty$ one arrives at the maximum population N_m. This generates the familiar curve of population growth termed the logistic or S-shaped curve. These equations are, of course, only very simple first approximations of growth.

In a particularly useful and insightful book by Hannon and Ruth (1997), the process of modeling is nicely developed with a computer program called STELLA.[46] In brief, this modeling approach starts by setting up a basic approach consisting of a set of variables and their interrelationships. More variables can then be added allowing the construction of quite complex dynamic models. To illustrate this consider the following scenario dealing with the process of logistic or sigmoid growth as represented by the above equations [5-1] to [5-8]. Consider a population of fish in a pond. This is considered as the first *state*[47] variable. In STELLA, this is known a *stock* or reservoir. The fish will be under a number of factors that will lead to an increase or decrease in their numbers. One of these factors is reproduction. These are *control* variables. These may be seen as inflows into the stock increasing its size or outflows decreasing its size (mortality or migration). Since we will need data (real or simulated) to drive the model, the next step is to determine the reproduction rate. This is a *transforming* variable called a *converter* in STELLA. We can now set this up as an equation. That is, the reproduction rate multiplied by stock equals the increase in fish over a selected discrete time interval. In other words:

$$N_{(t+1)} = \left[\left(\frac{\Delta N}{\Delta t} \right) * N_{(t)} \right] + N_{(t)},$$ [5-9]

where $\Delta N / \Delta t$ refers to the number of births per some time interval, which when added to the population, results in an increase in the fish stock.

---

[46] The software STELLA is available at: **http://www.hps-inc.com** or at (603) 643-9636.

[47] State variables (such as populations, fish stocks, information, prices, temperature, etc.) are major variables of importance that are acted on (changed) by other variables called flow or control variables.

At this point, the model simply reflects an increase in growth (exponential growth). Finally, we need to impose constraints on the fish population (their size cannot increase indefinitely in the confines of the pond). This is carried out in STELLA by setting up a feedback loop, which relates size of the population to the reproduction rate. In other words, the reproduction rate is dependent on some limiting value called the *carrying capacity*. As the population reaches the carrying capacity of the pond, the rate of reproduction drops significantly. There may also be increased mortality. This can be represented by an additional term, which includes the carrying capacity (K):

$$N_{(t+1)} = \left[\left(\frac{\Delta N}{\Delta t} \right) * N_{(t)} * \left(1 - \frac{N_{(t)}}{K} \right) \right] + N_{(t)}. \qquad [5\text{-}10]$$

Careful inspection of equations [5-10] and [5-7] will show that they represent the same thing. What remains now is to set up the equations that relate the variables and determine the model. Figure 5.2 illustrates the model generated with STELLA. Note that in Figure 5.2 there are three feedback loops. One represents the carrying capacity (K), the other the reproduction rate (R). The arrow from FISH to the circle also represents a feedback loop, in this case measuring the increase in offspring due to reproduction.

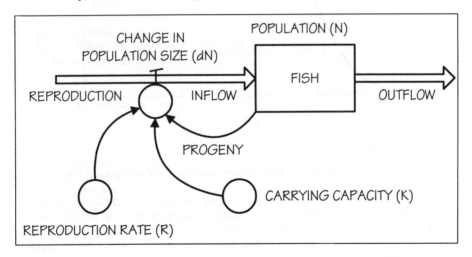

Figure 5.2. A simple model of fish population dynamics using STELLA.

We can now enter some values into STELLA, such as an initial population size of 10, a reproduction rate of 10% or 0.10 per month, and a feedback loop that reduces the reproduction rate as the carrying capacity of 375 fish is approached, and run the model. These values are, of course, arbitrary here, but need to be precise for modeling real

One of the advantages of modeling is the ability to simulate various conditions. This allows for continuous refinement of the model. This can be carried out by simply varying the values of the parameters in the model such the carrying capacity (K), the initial starting values, etc. In this way, one can model the dynamics of biological processes. The purpose of modeling is not prediction in the classical sense, but rather explanation. That is, to be able to demonstrate how the variables and the interrelationships among the variables produce the process.

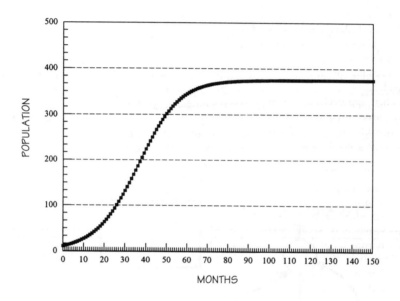

Figure 5.3. The logistic curve derived from the model using STELLA.

Models that are more realistic would include immigration and emigration as well as birth and death rates. Thus, even relatively simple models can rapidly become computationally complex. The interested student is highly encouraged to make use of STELLA and other such programs that allow the modeling of complex dynamic biological systems.

5.4.4. Other Approaches to Modeling

As mentioned earlier, there are numerous types of mathematical modeling approaches. A modeling technique that has become quite popular is computer simulation. Simulation means to *simulate* or artificially represent not everything in a system, but rather only certain details of interest (Carter and Huzan, 1973). Thus, a simulation is a dynamic *representation* of those aspects of interest to the investigator. Such a representation has also been called a model. It may be a physical model (*e.g.,* a scale model for wind tunnel flight testing) or an analog model such as a flight simulator for testing pilot flying skills. A third model is a mathematical model that simulates those system parameters of interest using equations. Any phenomenon that entails a set of interactions within a system of organizational entities lends itself to modeling. At least in theory, the simulation of biological processes via model-systems allows a means for gaining insight in the basic mechanisms of biological processes (Heinmetz, 1969). Such simulation models have been devised for population growth, species competition, genetic effects, evolutionary factors, demographic consequences such as migration, birth and death rates, etc. (Dyke and MacCluer, 1974). While simulations have been utilized for diverse purposes, they reflect common characteristics in that they focus on the interactions between variables using decision techniques. The literature on computer simulations is quite extensive and the interested reader is encouraged to peruse it. An especially readable account of simulation is the work by Casti (1997), which is highly recommended.

Another approach that also may use simulation techniques is derived from operations research (OR). Developed during WWII as *military operations research* and later shortened to OR, it was defined as the application of the scientific principles to the operation of organization involving people and resources (Dym and Ivey, 1980). The goal of OR is to arrive at optimized solutions. That is, maximizing one aspect while minimizing others. These may be maximizing profits while minimizing costs, or optimizing the number of doctors required at a hospital to address patient loads. Optimization often requires tradeoffs of costs versus benefits. Some of these concepts can be translated into a biological terms. Consider, for example, species competition in which coexistence is stabilized because each species maintains dominance within different environmental micro niches. Potentially, this could be modeled in terms of niche optimization. While these studies have led in a number of diverse directions, the potential of many of these experimental research areas has yet to be fully realized. Bartlett, in fact, questioned the utility of some of these studies, indicating:

> Such deductions are of theoretical interest, but their biological relevance is so dependent on the assumptions that very careful explanation and justification of these assumptions in necessary before we can take the conclusions seriously (Bartlett, 1968:208).

The rapid increases in computational power and the ability of storing large amounts of data, should facilitate developments in these potentially interesting areas and may rectify the earlier doubts espoused by Bartlett.

Thus, mathematical modeling is by no means a trivial endeavor. To master the art of modeling requires some background in matrix algebra, differential equations, Markov chains, game theory, linear programming (LP), probability theory, operations research (OR), graph theory, neural networks, fractals, chaos theory, artificial intelligence (AI), fuzzy logic and simulation, all areas than impinge on modeling but are beyond scope here. While perhaps intimidating, a biologist can acquire this background. Mathematical modeling then, can be viewed as an organizing principle that allows one to handle a large array of facts in a parsimonious manner (Beltrami, 1993).

5.4.5. The Challenge: Modeling the Development of Form

Two classical problems continue to confront workers. These are the development of form (morphogenesis) and the evolution of form (phylogenesis). These are related issues in the sense that both are time dependent processes and both produce changes in organismal form. As Edelman has succinctly stated:

> Thus the *particular* form of an animal in a species is the *combined* result of two factors: developmental regulation and natural selection affecting developmental dynamics (Italics in the original, Edelman, 1988:49)

Leaving phylogenetic questions aside, we need to ask how does a specific organismal form arise during morphogenesis? An answer to this question is not at all obvious and many of the details remain to be worked out. Partial answers have come, from genetics, from molecular biology and from embryology. All three are necessary ingredients that together facilitate the orderly development of the adult phenotype from a single cell—the zygote.

The fundamental problem of morphogenesis is then *pattern formation* according to Edelman (1988), and pattern formation implies organismal form. While some of the details are now becoming available, what remains to be demonstrated are the exact developmental mechanisms responsible, especially at points of discontinuity; for example, the initiation of tissue differentiation. That is, how is the pattern of differentiation *dynamically produced*, to use Edelman's terms? The solution to this question requires analysis in both spatial and temporal terms.

While it can be presumed that development (morphogenesis) is largely, if not totally, under genetic control, the details are not yet all that apparent. In terms of the answer to the question posed at the beginning; how does the organismal form arise; the influence of genetics—vis-à-vis DNA—while essential, is nevertheless, indirect. That is:

> It is important to notice, however, that while alterations in protein structure can give rise to enormous changes in the functioning of an animal or in particular traits, they do not *necessarily* form a *direct* basis for changes in shape at the scale of the animal or its organs or appendages (Italics in the original, Edelman, 1988:9)

This necessarily leads to the question of:

> How are cells of different types ordered in time or place during development to give species-specific tissue pattern and animal form? (Edelman, 1988:10).

In an endeavor to provide an answer to the above question, Edelman has proposed examining the driving forces leading to pattern formation. He identifies three such forces, which are cell division, cell movement and cell death. Each of which is under complex control of intracellular as well as molecular interactions that are external to the cell. Although the details involved are not fully understood (Edelman, 1988). Clearly, for a complete explanation of pattern formation, the identification of those genes that are significant during development need to be known as well as those molecules that regulate morphologic processes presumed to exist at the cell surface.[48] In addition, one needs to search for evidence of signals between cells and tissues that are actively involved in pattern formation.

Moreover, the genetic control is mediated by a sequence of regulatory events, which are, to a considerable extent, spatially determined. What this suggests is that form, *i.e.,* pattern formation, is not simply cell differentiation, but considerably more. It is the evolutionary basis of morphogenesis. In Edelman's words:

> The combination of patterns leading to form is the result of natural selection acting on developmental variants, the form and phenotypic function of which increase fitness (Edelman, 1988:130).

As stated earlier, the primary forces of cell division, cell and tissue movement and cell death, provide the driving forces for the development of form. The primary processes behind these driving forces are adhesion and place-dependent gene expression. According to Edelman:

> Adhesion leads to the formation of epithelia and ... allows for epithelial-mesenchymal transformation ... (These are involved) in a sequence of events that affect the genes that regulate adhesion as well as the genes that regulate cytodifferentiation (Edelman, 1988:131; material in parenthesis added).

It has become increasingly apparent that the cell surface is a major player in these events in that it mediates the signals from other cells and with other surfaces to form tissues. Involved here are morphoregulatory molecules which are expressed at the cell surface. The details are necessarily complex and remain ill-defined. Again, in Edelman's words:

> It is clear from this summary review that no strictly genetic account alone, not even one involving the detailed analysis of sequences of developmentally important genes, can provide a satisfactory answer to the developmental genetic question. It is equally clear that a purely mechanical or mechanochemical analysis that ignores this genetic control of developmental events is also inadequate (Edelman, 1988:132).

One approach that may be potentially successful is to view organismal growth as the on-going product of the interaction of millions of cells. Cellular differentiation aside, adult growth is a dynamic process ultimately involving net cell aggregates. Specifically, one can view net cell growth as the difference between mitotic divisions and cell deaths. While the actual details remain obscure, it may be possible to model the process. Moreover, while it may be argued that this represents a radically simplified view of a very complex process, it does allow growth to be treated as a population phenomenon for modeling purposes. What remains is to include form in the model. This may be achievable by setting up the following model as illustrated in Figure 5.4.

Figure 5.4 is an attempt to model individual growth as an aggregate of cells, which in totality account for form. A number of issues need to be overcome before this model is more than symbolic. These are identified with asterisks. The first one (form*) refers to

---

[48] Edelman calls these morphoregulatory molecules. They are responsible for the formation of sheets of cells (epithelium) and mesenchyme. Using chemical signals, these molecules also regulate the movements of cells.

the actual measurements that are needed to describe form. This will be taken up in detail in the next chapter. The second concern (limiting factor**) refers to the slowing down of growth as the adult stage is approached. This limiting factor, presumably, can be handled similarly to the carrying capacity shown Figure 5.2 and with equation [5-10].

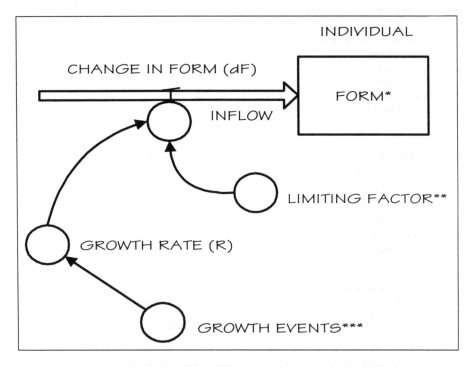

Figure 5.4. A potential model of form changes during growth.
(See text for explanation of starred items)

The last problem (growth events***) is intended to deal with changes that are time-dependent on one hand, and perhaps endocrine-influenced on the other, certainly the case for primates. Specifically for human data, this would involve the pubertal growth spurts. For other organisms, it could be instars,[49] etc. Modeling the presence of such events is complex. Previous approaches have tended to use regression as a means to fit curves to such growth data. However, regression approaches, useful as they may be, are not analytic and do not allow access to the underlying causes that are responsible for process.

---

[49] An instar is a stage of an insect or other arthropod between molts.

We now must leave these interesting issues dealing with the development of form from a morphogenesis standpoint, and return to the main theme of the volume: the quantification of form. Ultimately, it is the form that mediates the biological process, at all levels of scale. One only needs to consider how organismal form as modeled above, is influenced by the biological processes called growth. Thus, one of the consequences of morphogenesis is the continual alteration in form over time. How to precisely measure these changes remains a major challenge for biology.

Finally, it is argued here that without an adequate mathematical model for the representation of form, it may be premature to claim a precise understanding of the presumed underlying biological processes. Thus, what is needed is a formal model of form. The next chapter attempts to put current the representations of form into perspective. However, it is important to realize that, with respect to the study of form, the subject matter of this volume, there is no generally accepted theoretical foundation. That is, *a unified science of form* (or morphometrics) exists only in a very rudimentary sense.

KEY POINTS OF THE CHAPTER

This chapter illustrated some of the challenges involved in the development of a coherent model of form and the relationship of form to biological processes. The basis of complexity theory was introduced. It was also argued, based on complexity principles, that earlier equilibrium models are probably misleading. Such models are unable to adequately model biological processes such as morphogenesis, since they do not take into account aspects such as emergence, self-organization, etc. An attempt is made to emphasize that the view of the development of form must be made within a framework of complexity theory.

The notion of systems and system theory was briefly touched upon, a concept that remains under-utilized in the biological sciences. The development of modeling was then introduced. A number of types of models were discussed as well a simple example dealing with growth. The student was then exposed to one computer program called STELLA, which makes even the building of complex models relatively straightforward. Finally, the chapter ended with a return to the fundamental challenge facing biologists, which is to identify the biological processes, which are ultimately responsible to the changes visually observed in the biological form.

CHECK YOUR UNDERSTANDING

1. What is the difference between an ant and a collection of ants (an ant hill)? Is it just a matter of numbers or is something else involved?

2. What does it mean to say that complex systems are not in equilibrium?

3. What is meant by the expression "A systems view of the world"? Contrast the physical and biological science viewpoints.

4. List in a logical order the steps required to develop, test and validate a model.

5. What do the terms *static* and *dynamic* imply with respect to model building? How do they apply to a model of form?

6. Define the terms *state* and *control* variables. How are they related in the process of model building?

7. Describe what is meant by the expression "Out of chaos comes order". Illustrate with examples from the biological and zoological literature.

8. Describe the properties inherent in complex adaptive systems (*cas*). How are these manifested in the biological sciences? Provide some examples.

9. What is meant by the term *emergence* and how does it apply to complex biological systems?

10. What are some of the properties that make biological systems distinct from physical systems?

11. Why do biological systems have to be treated as "open systems"? How does the second law of thermodynamics apply to such open systems?

12. Why is a *feedback loop* essential in attempting to model biological processes such as growth?

13. Why is Newton's cosmology a good example of the systems approach?

REFERENCES CITED

Aris, R. (1978) *Mathematical Modelling Techniques*. San Francisco: Pitman Advanced Publishing System.

Bailey, N. T. J. (1967) *The Mathematical Approach to Biology and Medicine*. New York: John Wiley.

Bak, P. (1996) *How Nature Works*. New York: Copernicus, Springer-Verlag, Inc.

Bartlett, M. S. (1968) Biometry and theoretical biology. **In** *Haldane and Modern Biology*. Dronamraju, K. R. (Ed.). Baltimore: The John Hopkins Press.

Beltrami, E. (1993) *Mathematical Models in the Social and Biological Sciences*. London: Jones and Bartlett Publishers.

Blackith, R. E. and Reyment, R. A. (1971) *Multivariate Morphometrics*. New York: Academic Press.

Briggs, J. and Peat, F. D. (1989) *Turbulent Mirror*. New York: Harper and Row, Publishers.

Cambel, A. B. (1993) *Applied Chaos Theory: A Paradigm for Complexity*. New York: Academic Press.

Carter, L. R. and Huzan, E. (1973) *A Practical Approach to Computer Simulation in Business*. London: George Allen and Unwin Ltd.

Casti, J. L. (1997) *Would be Worlds*. John Wiley and Sons, Inc.

Clements, R. R. (1989) *Mathematical Modeling*. Cambridge: Cambridge University Press.

Cross, M. and Moscardini, A. O. (1985) *Learning the Art of Mathematical Modelling*. New York: John Wiley and Sons.

Dyke, B. and MacCluer, J. W. (1974) *Computer Simulation in Human Population Studies*. New York: Academic Press.

Dym, C. L. and Ivey, E. S. (1980) *Principles of Mathematical Modeling*. New York: Academic Press.

Edelman, G. M. (1988) *Topobiology*. New York: BasicBooks.

Hannon, B. and Ruth, M. (1997) *Modeling Dynamic Biological Systems*. New York: Springer-Verlag.

Heinmets, F. (1969) *Concepts and Models of Biomathematics*. New York: Marcel Dekker.

Holland, J. H. (1995) *Hidden Order*. New York: Addison-Wesley Pub. Co.

Holland, J. H. (1998) *Emergence*. Reading, Massachusetts: Perseus Books.

Kac, M. (1969) Some mathematical models in science. *Science* **166**:695-699.

Lestrel, P. E. (1997) *Fourier Descriptors and their Applications in Biology*. Cambridge: Cambridge University Press.

Lestrel, P. E. (1967) On some problems concerning adaptation in human populations. *Eugenics Quart.* **14**:155-156.

Levin, B. R. (1969) A model for selection in systems of species competition. **In** *Concepts and Models of Biomathematics*. Heinmets, F. (Ed.). New York: Marcel Dekker.

Li, C. C. (1955) *Population Genetics*. Chicago: University of Chicago Press.

Lotka, A. J. (1956) *Elements of Mathematical Biology*. New York: Dover Publications, Inc.

Mainzer, K. (1997) *Thinking in Complexity*. (3rd Ed.) New York: Springer-Verlag

Meyer, W. J. (1984) *Concepts of Mathematical Modeling*. New York: McGraw-Hill, Inc.

Reyment, R. A. (1991) *Multidimensional Paleobiology*. Oxford: Pergamon Press.

Saaty, T. L. and Alexander, J. M. (1982) *Thinking with Models*. New York: Pergamon Press.

Sheldrake, R. (1995) *A New Science of Life*. Rochester, Vermont: Park Street Press.

Singer, C. A (1959) *History of Scientific Ideas*. Oxford: Oxford University Press.

Thom, R. (1983) *Mathematical Models of Morphogenesis*. West Sussex: Ellis Horwood Ltd.

Waddington, C. H. (1957) *The Strategy of the Genes*. London: Allen and Unwin.

Waldrop, M. M. (1992) *Complexity*. New York: Simon & Schuster.

Weinberg, G. M. (1975) *An Introduction to General Systems Thinking*. New York: John Wiley and Sons.

Wilks, S. S. (1961) Some aspects of quantification in science. **In** *Quantification, A History of the Meaning of Measurement in the Natural and Social Sciences*. Woolf, H. (Ed.) New York: Bobbs-Merrill Co., Inc.

Zeeman, E. C. (1977) *Catastrophe Theory: Selected Papers 1972-77*. Reading, Mass. Addison-Wesley.

6. A FORMAL MODEL OF FORM

We are still in the phase of (an) explosive evolution in morphometric analysis, (with) a great diversity of techniques, and, more important still, (where a need for new) mental approaches to the subject is essential ...

Modified from Blackith and Reyment (1971:339)

6.1. INTRODUCTION

In the last chapter, a number of diverse issues that affect both morphogenetic and phylogenetic events were covered. At the conclusion of the last chapter, a very rudimentary model of the dynamics of organismal growth was briefly explored. That model is dynamic in that changes in form are viewed as a time-dependent process. However, that model is largely unverified at the moment, and therefore, can only be viewed in largely heuristic terms. The primary reason is that most measurements of the biological form remain inefficient and incomplete. In an attempt to remedy this, we now turn to a much more detailed analysis of a number of the representations of form. While the material here is largely static in character, in contrast to the last chapter, it marks the starting point for the actual measurement of the complex biological form.

6.2. A HEURISTIC MODEL OF FORM

Building a model to ultimately describe biological process must begin with measurement. Measurement, in a fundamental way, plays a major, if not always recognized, part in the elucidation of process. Two issues emerge here: [1] representation of form, this refers to the need for analytic procedures that can precisely measure forms and the differences between them and [2] the ultimate quantification of the biological process. The goal of a model of form is that it should act as a bridge between measurement and process. Moreover, it should shed light on the nature of the underlying biological processes, which are ultimately responsible for the form.

6.2.1. Justification for Quantitative Models of Form[50]

As indicated earlier (Section 1.2), without quantification there can be no science, at least no mature science. This is especially true for the physical sciences. Nevertheless, progress in quantification in the biological sciences; disciplines such as biology, paleontology, zoology, to name only a few, has been considerably slower in contrast to the physical sciences (physics, chemistry, astronomy, etc.). One reason for this state of affairs, as repeatedly mentioned, is that the objects under study in the biological sciences are simply more variable and complex. This has made it considerably more difficult to apply the quantitative methods that have proven so successful in the physical sciences.

---

[50] I am indebted to C. Wolfe, N. Garrett and D. Read for discussions of some of the material in this chapter.

Also, as alluded to earlier, with the rise of the natural sciences, the need for quantification of forms (biological and otherwise) was rarely contemplated. Quantification was simply not considered particularly relevant to the issues being addressed in the 18th and early 19th centuries. It is for these reasons that most of the biological sciences were initially (and as it has been argued to a considerable extent, still are) largely descriptive. Fauna and flora, for example, were described in terms of color, structural differences, etc. No attempt was made to quantify these attributes. Also mentioned earlier, another reason that may have initially acted to inhibit the use of numerical techniques, may have been the fact that the human brain, in concert with the highly developed visual system, readily generated descriptive information that was deemed appropriate for the investigator's purposes since it can be presumed that the questions then being raised were being adequately answered. It was only with the shift toward issues dealing with process, that this descriptive approach became insufficient.

Another more specific reason why a quantitative approach is to be preferred in the analysis of form, is as follows. A justification for focusing on the boundary of a form is that the human visual system treats boundary outlines as one of the features used for discrimination, followed by a focusing in on textural properties that lie within the form. That is, when we look at objects our visual system tends to initially lock on the boundary as a major feature. This is one justification for the need to precisely measure boundary information, not that the textural information within the boundary is any less important. Thus, from a visual perspective, it is the boundary that is viewed as the *shape attribute*. Thus, shape, together with the independent properties of size and texture make up some of the major attributes of form. Clearly, measurement of irregular boundaries has not been a trivial matter, and continues to remain difficult, if not impossible, to handle with simple metrics. Some of these issues will be taken up in detail below.

The human visual system, while highly sensitive to movement and color, is considerably less sensitive to the presence of small structural details in complex images or to subtle changes in the contours of outlines. When confronted with such irregular outline data, unless the differences are quite pronounced, the visual system can be overwhelmed, with the result that fine distinctions can be missed. This can be considered as another justification for the use of quantitative descriptions of form.

6.2.2. Developing a Concept of Form

As stated in the introduction, the idea of form is one of the most fundamental concepts underlying all of the sciences. In a fundamental sense, everything emanates from the notion of the displacement of space by an area or volume. If no space is displaced, there is no object, hence no form. Form is also dependent on *scale*, which can range from the very small (even at the atomic level) to the very large. Thus, studies as diverse as the classification of species, sediment particle shape, cell structure, diagnosis in pathology, and a host of other objects too numerous to mention, are all united in one way or another, by the analysis of form. All forms consist of a large number of shared aspects that include size, shape,[51] structure, etc. In the most basic sense, the human ability to readily discriminate forms by noting differences in color, size, shape, etc. is so well integrated

---

[51] In image processing and computer vision, the word morphology is synonymous with shape (Parker, 1997).

that the required behavior responses are largely unconscious. As mentioned in the introduction and reiterated here, of the five senses: sight, touch, hearing, smell, and speech; we probably rely most heavily on vision. However, all the senses come into play. In everyday routine activities, visual sense data tends to be acquired largely unconsciously and in qualitative and subjective terms.

Moreover, as already touched upon in Section 1.1.2, in contrast to artificial or man-made objects, naturally occurring phenomena generally exhibit considerable 'irregularities' or variability when one views individual members of a class of such objects. Thus, variability is of central importance in the natural world. Thus, all naturally occurring phenomena, at all scales, whether large or microscopic, will exhibit variability.

6.3. REPRESENTATIONS OF FORM

Morphometrics is a procedure, which facilitates the mapping of the visual information into a numerical (symbolic) representation (Read 1990). Various mappings have been proposed and their merits will be subsequently discussed. Three distinct representations can be recognized (Read and Lestrel, 1986; Lestrel, 1989). The first representation is dependent on *homologous-points* or landmarks and excludes the curvature between points. The second representation deals with the situation that arises when there are comparatively few landmarks present, or they are totally absent. In that case, the focus is on the *boundary-outline* of the form and landmarks play a secondary role. Clearly, these two representations focus on different aspects of the form. Both representations, homologous points and boundary-outline, have advantages as well as constraints. Approaches that combine the two representations, landmark and outline have been occasionally suggested, although little work is available (Read and Lestrel, 1986; Ray, 1990; Lestrel, 1997). Techniques that deal with *textural aspects* of form, both internal and external, characterize a third independent representation.

6.3.1. Specific Properties of Form

A number of elements play a role in the reliability and completeness of the representation being used to numerically describe the form (Lestrel, 1980; Mokhtarian and Mackworth, 1986). The following criteria are not necessarily complete or currently achievable with real data sets (Table 6.1). The last criterion in Table 6.1 merits further comment. The coordinate transformations identified as *uniform scaling, translation, rotation* and *reflection,* must each be invariant (Mokhtarian and Mackworth, 1986; Xu and Yang, 1990). These four transformations are further defined in Table 6.2.

The four properties in Table 6.2 share the characteristic that the form under consideration cannot change shape (be distorted in a visual sense) under each of these coordinate transformations. That is, these transformations must be *shape-preserving.* These aspects are termed here *coordinate* rather than *identity* because of the four properties, only three, translation, rotation and reflection, are 'identical' in the sense that both size and shape are maintained; that is, they remain unaltered after transformation. In the third case, uniform scaling; only shape remains invariant. In other words, only in the special case where both size and shape are 'identical' between comparisons, then, and only then, is the identity property or geometrical congruency satisfied. It is also important that the application of computational procedures that use these transformations, must not inadvertently cause alterations in these 'invariant' properties, otherwise computational results may become biased.

---

1. The representation must accurately describe the size and shape of the form as visually depicted,

2. The representation must be a unique solution for a given individual form,

3. The representation should be efficient in terms of computation; that is, if the computed variable set is reduced to a smaller sub-set, it must be with a minimal loss of accuracy,[52]

4. Large (global) aspects of the form as well as small ones (local) should be accurately mirrored in the values of the coefficients,

5. The representation provides for the extraction of a significant percentage of the visual information present,[53]

6. The representation should allow for the separate (uncorrelated) assessment of the components of size and shape (property of orthogonality),

7. Variables produced by the representation must provide for the *re-creation* of the form that is faithful to the original (observed) form,[54]

8. The representation must be invariant with respect to translation, rotation, reflection and scaling.

Table 6.1. Some properties of form representations.

Besides these four coordinate transformations, other uniform transformations such as the affine transformation exist. Here properties are relaxed with the effect that the shape is altered. As a simple example of this transformation, consider a square turned into a parallelogram. The key feature here is that while distortion is introduced, parallel lines defining the boundary of the square *remain* parallel in the parallelogram. This mapping can still be considered as uniform, or linear, in the sense that proportions are maintained and that lines remain straight and parallel to each other, only angularity is not being preserved.

Finally, non-uniform transformations exist, such as those produced by the plasticity or flow of material. This refers to *elasticity* in an engineering sense. Here, points

---

[52] For example, if a curve-fitting procedure is employed to numerically describe the size and shape of the boundary of a form, then it is incumbent upon the investigator to provide a measure of the accuracy of the curve-fit. This is usually the residual or difference between the observed form and its predicted curve-fit (Chapter 9).

[53] What is a 'significant percentage' remains a largely unresolved issue at the moment. Nevertheless, the use of an inefficient method such as CMA tends to insure that only a very small percentage of the morphological information present is being measured (Chapter 7).

[54] This refers to what is called the *information preserving* property characteristic of both boundary morphometrics (Chapter 9) and, in part, structural morphometrics (Chapter 10).

embedded in such a material are allowed to move in a way which causes a stretching or compression of the relationships of these points to each other (Bookstein, 1991; Reyment, 1991). The effects of such transformations are of special interest and lead to biorthogonal grids (Section 8.3) and thin-plate splines (Section 8.5).

---

1. **Translation**—each point (x, y, z) is moved a constant, Δx, distance in the x-direction, a constant, Δy, distance in the y-direction and a constant, Δz, distance in the z-direction. These translations within the Cartesian plane are parallel, and orthogonal with respect to each plane. Each form maintains its size and shape as it is being shifted to a new location in the coordinate system. This can be termed as a one-to-one mapping of the whole form with itself.

2. **Rotation**—each point (x, y, z) is moved through a constant, λ_{xy}, angle in the xy plane, a constant, λ_{xz}, angle in the xz plane and a constant, λ_{zy}, angle in the zy plane. Each of these rotations is in relation to a common origin. This rotation can be independently carried out through all three angles. Again, size and shape has to be maintained. This is also a one-to-one mapping of the form with itself.

3. **Reflection**—each point (x, y, z) reflects sign changes in an orderly way so that a mirror-image is produced. Each form is again invariant with size and shape as it is being symmetrically flipped to a new location in the coordinate system. It is still a one-to-one mapping of the whole form with its reflection (This property has been applied in Section 9.3.4).

4. **Uniform scaling**—each point (x, y, z) is scaled by some constant factor (often a multiplier, *e.g.*, based on a distance, area or perimeter) in such a fashion that shape is not distorted. Again, by making the scaling factor equal to 1.0, the form will revert to a one-to-one mapping with itself.

Table 6.2. Coordinate transformations.

6.3.2. Size, Shape and Structural Considerations

All forms consist of a large number of shared aspects that include size, shape, color, structure, patterning, etc. However, the numerical characterization of such forms has been a more challenging endeavors than may be at first realized. This is especially the case in morphological studies of biological organisms. Moreover, even such concepts as size, shape and form have not been without controversy. For example, form and shape have been used interchangeably. According to Webster's Unabridged Dictionary (1983), form is defined as "the shape or outline of anything; figure; image; structure, excluding color, texture and density". Upon looking up shape we find that things are hopelessly muddled with shape defined as "...outline or external surface" or "the form characteristic of a particular person or thing". According to these definitions, 'shape' and 'form' are interchangeable, to be viewed as identical. It is of historical interest that the great biologist D'Arcy Thompson also viewed form and shape interchangeably. Clearly, this unsatisfactory state of affairs is in need of re-definition if the concept of form and its attributes are to be precisely defined. Precise definition, after all, is one of the hallmarks of all sciences.

To alleviate some of these problems, a simple linear formulation was proposed (Needham, 1950; Penrose, 1954) as:

$$\textbf{Form = Size + Shape.} \qquad\qquad [6\text{-}1]$$

While both size and shape are biologically important attributes of form, their contributions differ. Moreover, the presence of large size differences can act to confound or swamp shape differences, hence the use of various normalization procedures.

Size can be defined as a quantity that depends upon dimensional space. This can be made clearer by focusing on size-differences. In a one-dimensional world, differences in size can be viewed as differences in vector length. In two dimensions, linear measurements in combinations (such as ratios) have proven to be inadequate, and area becomes one definition of size. The perimeter has also been utilized (van Otterloo, 1991) but the use of perimeter, at least in a biological context, cannot be recommended here as a size measure, because very convoluted outlines would have very large values while smooth outlines would have small values. Area, it is argued here, is the preferred standardization for 2-D outlines. In 3-D, volume would be the appropriate size measure.

Shape, in contrast, is a quantity that has been traditionally difficult to adequately define. It has been characterized as 'residual', or what is left after size has been controlled for. Various attempts have been made to try to quantify shape using relatively simple measures. These have been termed shape factors, but cannot be considered satisfactory except for the simplest of shapes (Moore, 1982; O'Higgins, 1997).

A more technical approach to shape implies that it remains invariant under scaling, translation, rotation and reflection (van Otterloo, 1991; Lele, 1991). While accepting this latter definition, one also needs stress the visual aspect, which leads one to *define shape in terms of the boundary of the form*. Thus, when appropriate, one can control, or normalize for size, by making the bounded area of 2-D forms equal for all comparisons (for 3-D the normalization would be based on volume). This procedure effectively diminishes the influence of size close to zero. Then, and only then, one can equate 'form with shape' and re-write equation [6-1] above to:

$$\textbf{Form = Shape.} \qquad\qquad [6\text{-}2]$$

Technically, this implies that only if the shapes are identical after scaling for size, are they then congruent. Moreover, only when conditions are specifically defined, as here, can one begin to legitimately equate 'form' with 'shape'. To do so otherwise only engenders confusion.

The earlier attempt to define form, equation [6-1], was subsequently extended to include structure:

$$\text{Form} = \text{Size} + \text{Shape} + \text{Structure}. \qquad [6\text{-}3]$$

Definitions of the aspects of size and shape have been alluded to already. Structure, can be defined as a quantity that describes the 'within boundary' aspects. It can refer to both internal and external considerations. For example, the external 'roughness' on the outside of an object, or *e.g.*, the internal patterning or 'orientation' of the cancellous bony spicules (trabeculae) that make up the femur head. Techniques for the analysis of structure include coherent optical processing, Fourier transforms and wavelets, which are taken up in Chapter 10 (Lestrel, 1980, Lestrel, 1997).

6.3.3. A More Realistic Model of Form

While the above linear formulation, Form = Size + Shape + Structure, was initially considered as a step towards a formal model of form, it only represents a fraction of the attributes that define all forms, especially those found in nature. Some of these attributes are illustrated with Figure 6.1.

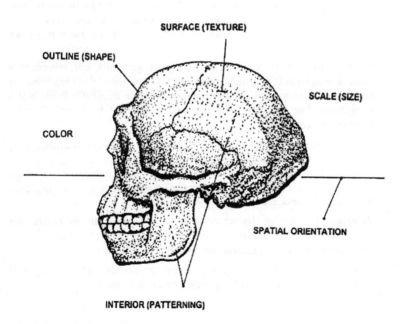

Figure 6.1. Some morphological attributes of form.

Clearly, the lagging developments leading to such a theoretical model of form are due to its formidable difficulties. These include: [1] problems of definition; that is, a vocabulary that precisely conveys information about form, [2] problems of description, in quantitative terms, of the various attributes, and the [3] global: that is, the inclusion of the set of attributes into a useful multivariate model. It seems apparent that, at the moment, there is no 'single' numerical value, which can adequately describe a form in all its aspects. Table 6.3 represents a first attempt to rectify some of these issues.

---

1. *State*—this attribute defines whether it is a solid, liquid or gas.[55] A three-item vector may be applicable to indicate percentages of each quantity. For purposes here, it is considered as being equal to a solid. [1.0, 0.0, 0.0].

2. *Size*—this is a measure of spatial dimension(s) and can be viewed as a displacement in space such as: [1] length in 1-space, [2] area in 2-space and [3] volume in 3-space.

3. *Shape*—this is a boundary phenomenon. It refers to the boundary outline of a form in 2-D or 3-D. It is a geometric property, as defined here, and not related to the notion of surface as such (see *Surface*). Its focus is on curvature. In 1-space, all point forms and all line forms have the same shapes.

4. *Orientation*—this refers to location in space, generally with respect to a reference system (Cartesian, polar, etc.). Orientation, for simplicity, is considered here as static at any point in time for a specific form. Note: Coordinate-free morphometric methods would not be constrained by orientation requirements.[56]

5. *Surface*—this is characterized as having zero thickness and consisting of two primary properties: *texture* and *color*. Textural properties are bounded by the shape; that is, they lie within the boundary outline. Surface texture refers to the presence of vascularities, smoothness, pits, bumps, roughness, etc.

6. *Interior*—this is also characterized by textural and color properties but has thickness in contrast to surface. It can be usefully viewed as orthogonal or normal ($\perp$) everywhere to the surface of the form.

7. *Substance*—this is based on physical properties such as thermal expansion, mass or weight, density, conduction, modulus of elasticity, hardness, etc.

Table 6.3. Attributes of a more realistic model of form.

---

[55] It is admitted that a 'gaseous' or 'liquid' state may be largely meaningless in the context of form as used here.
[56] Under certain circumstances, orientation in space may be important and need to be retained. This property is, however, lost with coordinate-free methods (Chapter 8).

The equations of form discussed above, such as [6-1] or [6-3], and set up in terms of the linear equations, can only be considered among the simplest of relationships. Moreover, there is considerable skepticism regarding the linearity of these simple models. Actually, many more attributes make up form. The morphological attributes shown in Figure 6.1 can also be set up as a formal model:

Form = F = <state, size, shape, orientation, surface, interior, substance> [6-4]

In contrast to the other earlier proposed models, note that the notation[57] in this model makes no assumptions about linearity (Lestrel, 1997). Each of these seven attributes (each of which may be further composed of multiple properties) is listed in Table 6.3. These morphological aspects, as mathematical constructs, are intended to be dimensionally independent attributes. That is, with the correlation between attributes equal to zero. In mathematical terms, this independence is termed *orthogonality*. Besides the property of perpendicularity in a geometric sense, orthogonality also implies that the dot product of two vectors is equal to zero (Strang, 1994). Orthogonality is an important property, which plays a prominent part, for example, in Fourier analysis (Chapters 9). Implicit in Table 6.3 is not only the application of geometric constructs but also the need for simplification; otherwise, the quantification of complex biological forms can rapidly become intractable. Thus, the ability to describe real objects with modeling requires, for example, the use of points and lines with infinitesimal diameters and thickness, while the definition of surface having no thickness, is another.

6.3.4. A Dynamic Model of Form

As stated above, modeling requires simplification. An example of a simplified model is a road map. It provides just enough information to accomplish the goal; namely, providing *directions*. Map details that are not specifically associated with that goal are generally excluded. Also involved is the issue of scaling as maps can only provide a limited amount of detail. As Holland (1998) has cogently pointed out, one way to isolate those aspects of interest (goals) is to derive functions. Because these functions represent one-to-one mappings, they provide a way of eliminating detail. Holland then utilized game theory concepts such as the notion of *states*. States (*e.g.*, the arrangement of chess pieces at any point in of a game) can be considered as a summary or condition at a particular point in time. The *state space* is then the collection of all such arrangements.[58] This set of all arrangements (state space) is determined by successive *moves*. One can then build a 'tree' of such moves. The *root* of the tree is the *initial state*; branches lead to subsequent states, with leaves becoming *end states*. The number of leaves of such trees (actually bushes) can quickly grow very rapidly. Thus, a comparatively small number of rules (functions) can define an enormous number of states. In real games, of course, one uses *strategies* in an attempt to 'win'. Strategies represent decision-making responses to

---

[57] This notation is formally known as an n-tuple, here a 7-tuple. While perhaps unfamiliar to the reader, the concept is an extension of such ideas as a triplet (a 3-tuple) such as Cartesian coordinates (x, y, z). The notation is widely used in such fields as computer science and computational linguistics. I am particularly indebted to C. Wolfe for his efforts in developing this modeling approach.

[58] Not the total set of possible moves, but only a subset, those actually allowed by the 'rules' of the game.

ones opponent. Such decision-making may be good or bad. Game theory provides a *complete strategy* by prescribing a move for each state that can be encountered. Here also functions can be constructed:

> ... that define the correspondence between game states and the moves prescribed by strategy ... For every strategy a function exists that describes that strategy (Holland, 1998:39).

With this brief background, we can now shift to the construction of *dynamic models*. The ultimate purpose of such dynamic models here is to try to detect the presence of 'laws' underlying biological processes. Laws, in this sense used here, *correspond to the rules of a game*. These *laws of change*, to use Holland's terms, specify the succession of states, for instance, in a biological context, changes in the human form during growth at 4 years, 8 years, 12 years, and so on. Again, scale is involved, in Holland words:

> To build a dynamic model we have to select a level of detail that is useful and then capture the laws of change at that level of detail. ... It may be quite difficult to construct a detailed model that is "faithful" to the system being modeled (Holland, 1998:46).

Figure 6.2 (Adapted from Holland, 1998) views the modeling techniques offered in the last chapter from a somewhat different perspective. The attempt here is to integrate the dynamic model with the measurement of form as derived from morphometrics. Accordingly, the laws of change alluded to above, can now be precisely defined with the use of *transition functions*.[59] These are functions that assign values from one state to the next. These transition functions are often set up in terms of differential or difference equations as already encountered in Chapter 5 (see Section 5.4.3 for a simple example).

Biological growth is influenced by 'internal' events as well as 'external' ones. These can be modeled as inputs so that according to Holland:

> ... the transition function provides a correspondence between each possible (state, input) pair and the state that results (Holland, 1998:48).

Figure 6.2, top, displays the real world (form) as a mapping from $t \rightarrow t + 1$ (a *state space* change). The circle and ellipse represent the *observed* biological form changing from $t \rightarrow t + 1$. This change in form, from a circle to an ellipse, symbolizes changes in size, shape, structure, etc., of the form over time. In contrast, Figure 6.2, bottom, displays the change in form as a dynamic representation or model, using transition functions (algorithms) to model the change in form. Now the change in form, from a square to a rectangle, symbolizes the algorithmic changes in size, shape, etc., of the form from $t \rightarrow t + 1$ (a *state space* transformation). The approach here is very briefly sketched, simply intended to assist in visualizing the dynamic modeling approach initially presented in Chapter 5. Nevertheless, it may provide a basis for building actual models of form. However, what remains to be developed are the algorithms used to represent the transition functions, which will clearly be a challenge for the 21st century.

---

[59] A transition function may have to be expressed in stochastic terms; in which case, these are set up as transition probabilities of moving from one state to another.

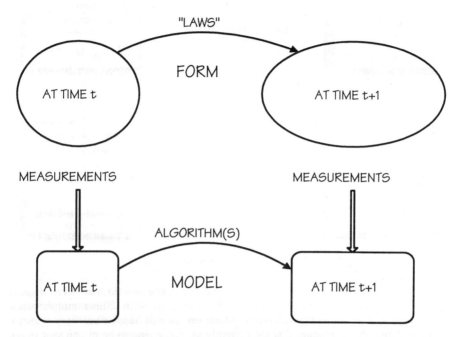

Figure 6.2. Transition functions and states.

If one considers the growth of organisms to be one of the biological processes of fundamental importance, then morphometrics plays a key role. Thus, considerable emphasis has been placed by numerous investigators on the development and refinement of morphometric methods for the numerical characterization of form. The central challenge of morphometrics lies in the capture of the visual information that resides in all forms and to re-express it in numerical terms to facilitate further analysis. That is, what is required is the ability to capture a significant amount of the visual information that is present. In contrast to the human capability of rapidly identifying and classifying this visual information, the mathematical description of the content of these visual images has been painfully slow (Lestrel, 1997). These considerations reinforce the need for a mathematical representation or model of form. However, there are substantial issues that remain to be addressed in the course of model development. These primarily deal with the creation of proper algorithms that accurately reflect the changes in form.

Again, in an effort to make the modeling process clearer, examine Figure 6.3. This figure is similar to Figure 1.1 but attempts to depict the modeling process in algorithmic terms. Clearly, the challenge that presents itself is less in the development of a model of form than in its implementation. Implementation will require the development of specific algorithms that describe the form and allow for testing, analysis and verification of the model being used to describe biological processes such as growth. In other words, what is needed is an *algebra of form*.

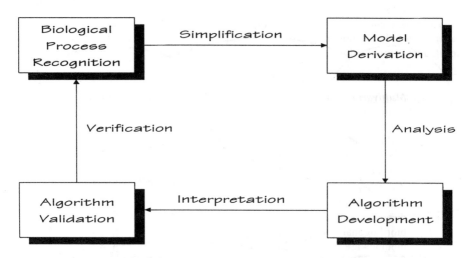

Figure 6.3. A procedural approach to modeling.

The next section introduces some of the numerous attempts that have been made to generate mathematical representations of the biological form. These morphometric procedures, as a group, are quite diverse. Moreover, as will become apparent in Part II of this volume, the techniques that are currently available tend to be insufficient in one way or another.

6.4. MORPHOMETRIC PROCEDURES

Measurement is not an inconsequential issue, and a number of unresolved problems exist. These problems continue to be present because of a long-held but unwarranted assumption that the measurement procedures applied in the biological sciences are straightforward, easily invoked and therefore requiring little consideration. It is partially in response to this assumption that new and more sophisticated approaches have emerged.

Consequently, morphometrics, as viewed here, consists of a considerably broadened definition, which encompasses, at this moment, a number of separate and distinct approaches that are intended to deal with the numerical description of form at all levels of scaling (Table 6.4). Ultimately, these diverse approaches will need to be merged into a coherent and unified model. Some of these morphometric methods used to analyze form can be placed into the three broad categories: [1] *homologous-point* methods, [2] *boundary-outline* methods and [3] *structural* methods.

One of the strengths of homologous-point methods is that they allow for the localization of features at the expense of global aspects. Boundary outline methods, on the other hand, are largely global in nature, although localized aspects, at times, can be identified. These two approaches can be subdivided further into four groups as listed in Table 6.4 and Figure 6.4, which in turn, contain specific techniques that have been applied to data sets (Lestrel, 1997). Note: While the four-fold division (Table 6.4) may be useful as a classificatory device, it must be emphasized that with respect to the two latter entries; that is, boundary and structural morphometrics, considerable overlap of the techniques is involved.

---

1. ***Multivariate morphometrics*** (Chapter 7) — This is the earliest application of morphometric techniques to obtain data. It refers to the utilization of multivariate statistical methods to determine whether within- and between-groups differences exist. Generally applied to data sets composed of distances, angles and ratios (Blackith and Reyment, 1971; Reyment, *et al.*, 1984).

2. ***Coordinate morphometrics*** (Chapter 8) — These are landmark-based representations. They include the conventional metrical approach of distances and angles, Euclidean distance matrix analysis, as well as biorthogonal grids, finite elements and thin plate splines. The latter three methods are distinctly different in that they focus on deformations. All five methods can be viewed as *coordinate-free*; that is, they are independent of the coordinate system.

3. ***Boundary morphometrics*** (Chapter 9) — These are boundary-based representations and include median axis functions, Fourier descriptors, eigenshape analysis, elliptical Fourier analysis and wavelets. These methods are *information preserving*; that is, the original form can be recreated at will, and lastly.

4. ***Structural morphometrics*** (Chapter 10) — These are texturally based representations (external and/or internal) and include Fourier transforms, coherent optical processing and wavelets. These techniques, with some exceptions, are also *information preserving*.

Table 6.4. The four major divisions of morphometrics.

Morphometrics then, can be viewed as a set of procedures, which facilitate the mapping of the visual information of form into a mathematical representation. Currently, morphometrics represents the combination of four diverse approaches, each with a number of distinct methods for dealing with the quantitative analysis of morphology.

Each method utilizes data sets in unique ways and focuses on different aspects of the form. Moreover, these methods tend to be viewed as largely independent of each other because of the lack of a formal unifying model. These are, by no means the only methods to date, but represent some those that have received the most attention over the last three decades. Each of these methods will be briefly taken up in the final four Chapters.

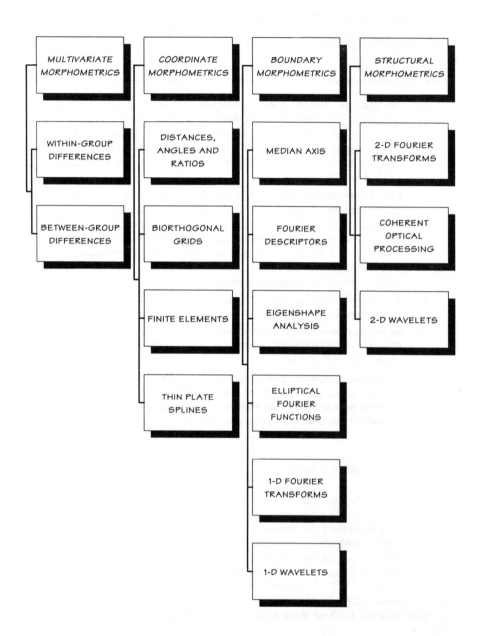

Figure 6.4. A schematic flowchart of morphometric procedures.

KEY POINTS OF THE CHAPTER

This chapter continued the elaboration of the challenges involved in the development of a coherent model of form and the relationship of form to biological processes. The emphasis here returned to measurement as the central focus. The chapter started with a heuristic model of form and led to a number of specific properties that were presumed to be inherent in all forms. Form was initially set up as a simple linear equation and eventually transformed into a complex but more realistic formulation that did not require linearity. While the model of form developed here was initially essentially static, it does lend itself for modeling. A model was eventually proposed which allowed for a dynamic analysis of the biological form based on state variables and transition functions.

The latter part of the chapter marked a return to the core of the volume; namely, morphometrics. Four major divisions were identified and each of these was divided into subdivisions that reflected some of the different morphometric techniques currently in vogue.

CHECK YOUR UNDERSTANDING

1. Consider two of the basic properties of form; size and shape. Why is it easier to measure size in contrast to shape?

2. If each of the four coordinate transformations were relaxed, one at a time, what effect would it have on the biological form?

3. Besides the six attributes of form illustrated with Figure 6.1, are there others that could or need to be considered?

4. Is color an important attribute? Consider some examples from the biological, botanical, and zoological literature that demonstrate the importance of color within an ontogenetic or phylogenetic context.

5. How would you numerically describe the attribute of color as seen in the biological form? What difficulties could be encountered in the process?

6. Distinguish between *boundary, coordinate*, and *structural* approaches; why are none of these approaches complete in themselves?

7. What are some of the deficiencies inherent with the conventional metrical approach consisting of distances, angles and ratios?

8. Why is the property of orthogonality essential in considering the attributes that comprise the model of form defined with equation [6-4]?

9. How can one balance the need for simplification in model building with the desire to create a realistic model that describes the biological process? Hint: Consider growth as an example of such a process.

REFERENCES CITED

Blackith, R. E. and Reyment, R. A. (1971) *Multivariate Morphometrics*. New York: Academic Press.

Bookstein, F. L. (1991) *Morphometric Tools for Landmark Data*. Cambridge: Cambridge University Press.

Holland, J. H. (1998) *Emergence*. Reading, Massachusetts: Perseus Books.

Lele, S. (1991) Some comments on coordinate free and scale invariant methods in morphometrics. *Am. J. Phys. Anthrop.* **85**:407-417.

Lestrel, P. E. (1980) A quantitative approach to skeletal morphology: Fourier analysis. *Soc. Photo. Inst. Engrs. (SPIE)*, **166**:80-93.

Lestrel, P. E. (1989) Some approaches toward the mathematical modeling of the craniofacial complex. *J. Craniofacial Genet. Dev. Biol.* **9**:77-91.

Lestrel, P. E. (1997) *Fourier Descriptors and their Applications in Biology*. Cambridge: Cambridge University Press.

Mokhtarian, F. and Mackworth, A. (1986) Scale-based description and recognition of planar curves and two-dimensional shapes. *IEEE Trans. Pattern Anal. Mach. Intell.* PAMI-**8**:34-43.

Moore, W. J. (1982) Skull form in hominoids. In *Progress in Anatomy, Vol. 2* (Harrison, R. J. and Navaratnam, V. (Eds.). Cambridge: Cambridge University Press.

Needham, A. E. (1950) The form-transformation of the abdomen of the female pea-crab, *Pinnotheres pisum*: *Proc. Roy. Soc. (Lond.)* **137**:115-136.

O'Higgins, P. (1997) Methodological issues in the description of forms. In *Fourier Descriptors and their Applications in Biology*. Lestrel, P. E. (Ed.). Cambridge: Cambridge University Press.

Parker, J. R. (1997) *Algorithms for Image Processing and Computer Vision*. New York: John Wiley and Sons.

Penrose, L. S. (1954) Distance, size and shape. *Ann. Eugen.* **18**:337-343.

Ray, T. S. (1990) Application of eigenshape analysis to second order leaf shape ontogeny in *Syngonium podophyllum* (Araceae). In *Proceedings of the Michigan Morphometric Workshop*. Rohlf, F. J. and Bookstein, F. L. (Eds.) University of Michigan Museum of Zoology Special Pub. No 2:201-213.

Read, D. W. and Lestrel, P. E. (1986) A comment upon the uses of homologous-point measures in systematics: A reply to Bookstein *et al. Syst. Zool.* **33**:241-53.

Read, D. W. (1990) From multivariate to qualitative measurement: Representation of shape. *Hum. Evol.* **5**:417-429.

Reyment, R. A. (1991) *Multidimensional Paleobiology*. Oxford: Pergamon Press.

Reyment, R. A., Blackith, R. E. and Campbell, N. A. (1984) *Multivariate Morphometrics (2nd Ed.)*. New York: Academic Press.

Reyment, R. A. (1991) *Multidimensional Paleobiology*. Oxford: Pergamon Press.

Strang, G., (1994) Wavelets. *Amer. Sci.* **82**:250-255.

Thompson, D. W. (1942) *On Growth and form. (2nd Ed.)* Cambridge: Cambridge University Press.

van Otterloo, P. J. (1991) *A Contour-oriented Approach to Shape Analysis.* New York: Prentice-Hall.

Webster, N. (1983) *New Universal Unabridged Dictionary* (2nd Ed.). Cleveland: Simon and Schuster.

Xu, J. and Yang, Y-H. (1990) Generalized multidimensional orthogonal polynomials with applications to shape analysis. *IEEE Trans. Pattern Anal. Mach. Int..* 12:906-913.

PART TWO:
MORPHOMETRIC TECHNIQUES

7. MULTIVARIATE MORPHOMETRICS

Et harum scientarum porta et clavis est Mathematica.
Mathematics is the door and key to the sciences.

Roger Bacon (*ca.* 1214-1294)

7.1. INTRODUCTION

The application of multivariate statistics in the biological sciences, and specifically morphometrics, can already be seen in the initial developments that gave rise to the various multivariate statistical techniques in common use today. The focus of these methods was aimed at elucidating statistical properties of within- and between-groups encountered with real biological data sets. Nevertheless, the formal recognition of such statistical applications, especially in the biological disciplines, did not really b3come widespread until the 1970's. The appearance of *Multivariate Morphometrics* by Blackith and Reyment (1971) and *Numerical Taxonomy* by Sneath and Sokal (1973) did much to focus attention on the relationship between multivariate statistics and morphometrics. The former book was re-issued in abridged form in 1984 as a second edition (Reyment *et al.,* 1984). Another more recent volume by Reyment (1991), represents a continuation of those efforts and can be profitably consulted. Finally, the book by Davis (1986) is a particularly useful and readable account of the geometric interpretation of multivariate statistical methods. The following material will be limited to multivariate statistical techniques that are specifically applicable in the morphometric context.

7.1.1. Historical Background

In a peculiar way, one of the earliest relationships between multivariate statistics and morphometrics can be traced to the work of Alphonse Bertillon (1853-1914) who developed a system for criminal identification used by the Paris police. Bertillon, whose research stimulated the interest of Francis Galton (discussed in Section 4.2.3), came from an eminent French statistical family (Porter, 1986). The *Bertillon* system (1882) incorporated a number of body measurements, the height and length of finger, arm and foot, as well as a physical description, and photographs. The Bertillon system was eventually superseded by fingerprinting. It is with the attempt to look at body measurements simultaneously, that Bertillon is anticipating multivariate analysis.

However, the real beginnings start with the first decades of the 20th century, especially with the work of Karl Pearson (Pearson, 1901; 1926). Although frequent application of these techniques did not take place until the 1960's, many of the actual statistical methods employed were developed earlier. These include principal components (Pearson, 1901; Hotelling, 1933), the multivariate generalization of the t-statistic, the Hotelling T^2 (Hotelling, 1931), discriminant functions (Fisher, 1936), the generalized distance or D^2 statistic (Mahalanobis, 1936) and factor analysis (Spearman,

1903; Cattell, 1965a; 1965b). Multivariate morphometrics evolved to meet the need for procedures aimed at measuring the degree of similarity within and between two or more forms using multiple measurements. This led to the development of techniques that facilitated statistical decisions based on 'similarity criteria', which allowed the classification of similar objects into categories (Spoehr and Lehmkuhle, 1982; Nadler and Smith, 1993). These events led, quite early, toward development of multivariate statistics and subsequently to the application of such methods to morphometric data. Application of these methods was predicated on the grounds that the simultaneous utilization of numerous variables may provide more information than a large number of individual variables being analyzed separately.

7.1.2. Multivariate Procedures

Multivariate analysis, viewed in a statistical sense, has been defined as: the use of quantitative methods to discover the structure of interrelationships of multiple measurements (Bernstein, 1988; Affifi and Clark, 1984). Here, the statistical approach of multivariate analysis is brought to bear on the data of morphometrics. In 1985, Reyment defined multivariate morphometrics:

> ... to be the application of methods and principles of multivariate statistical analysis to the study of variation in plants and animals (Reyment, 1985:591).

It is argued here that new developments, in concert with a new broader outlook, may make the above definition unduly narrow. However, the importance of multivariate analysis as a statistical tool in a morphometric context cannot be overemphasized.

In contrast to the other procedures described in Chapters 8, 9 and 10, and as noted earlier, multivariate morphometrics consists of a set of procedures designed to elucidate within- and between-group relationships with the focus on the morphological form. These methods can be extremely useful for gaining insights about the data in terms of: [1] how the variables are structured and [2] how the groups are related. Many of these techniques are based on linear (first order) models that use data sets composed of multiple original variables used simultaneously.

One implicit assumption has been that since morphological forms were complex, one simply needed to take more measurements. The idea behind that approach was (and still is) that it should be possible to recover the essential elements of the form with the use of many individual measurements. Nevertheless, two obstacles initially mitigated against the widespread use of these multivariate statististical methods: [1] the extensive and complex calculations required and [2] the problem of correlations between multiple measurements. Some of these difficulties (but not all) have been gradually alleviated with the increasing use of computers. Computers eased the computational burdens and certain developments in statistics, led to techniques that 'de-correlated' the data sets. Such computer programs began to be routinely available on mainframes from the 1970's on. Since the middle 1980's, they have also been available on personal computers as well. The mainframe packages commonly cited in the literature are BMDP, SAS and SPSS (Affifi and Clark, 1984; Bernstein, 1988). Packages utilized in the PC environment now include STATISTICA, SYSTAT, SIGMASTAT, STATGRAPHICS, NCSS, etc., to name a few. This list is by no means exhaustive and others could have been mentioned. These are listed only because of their familiarity. All these packages contain most of the necessary routines required to carry out analyses of multivariate morphometric data sets.

The overwhelming numbers of studies using multivariate morphometrics remain based on a measurement system composed of distances, angles and ratios (CMA). These measures are taken up in some detail in Chapter 8. Whether these measurements are the most appropriate for the numerical description and analysis of the biological data sets is seldom considered. Further, attempts to define size and shape using variables derived from CMA, have led to various models (Penrose, 1954; Mosimann, 1970; Sprent, 1972); yet, how such approaches can be related to visual notions of shape (*e.g.*, the outline of form) remains questionable. While a useful discussion of the relationship of multivariate morphometrics to shape analysis can be found in Reyment (1985), there is scarcely any comment of the applicability of these measures serving as data. Moreover, the approach using CMA, as stated earlier, remains particularly inefficient if one is interested in quantifying the visual information expressed by the form boundary (Chapter 9) or textural considerations within the form (Chapter 10).

Nevertheless in a basic sense, multivariate morphometrics can be viewed as an extension or adjunct to coordinate morphometrics (Chapter 8), in that the conventional metrical approach (CMA) consisting of distances angles and ratios, most often forms the data sets needed for the application of the various multivariate techniques.

Multivariate morphometrics is treated here as composed of a series of sophisticated mathematical approaches that deal with a variety of specific morphometric problems. Some of these applications include: [1] establishing the *closeness of relationship* or *similarity* between different forms, [2] measuring the variation that is present, using a set of *uncorrelated* variables, [3] investigating the *structure* of the measurements used to describe the form and [4] identifying the components of size and shape. Finally, because the number of variables of interest often exceed three; that is, requiring more than three dimensions to be simultaneously considered, there is the additional problem of visualizing them in higher dimensional space. This is often cleverly addressed by graphical methods. While the technical details underlying the computations of these methods are beyond the scope of this chapter, a few comments may be useful.

For computational purposes, the data set is set up as a matrix. Such a matrix is defined here as a rectangular array of numbers composed of n-rows and p-columns where the variables form columns and the individual specimens (or cases) measured form rows. The comparison of columns, *i.e.*, *variables*, can be considered as a within-group procedure, while the comparison of rows, *i.e.*, *individual* cases, is defined as a between-group procedure (Mardia, *et al.*, 1979). The original variables are often transformed in a specific manner and then re-expressed into a smaller set of new, independent (orthogonal) variables that can be embedded in a set of rectangular coordinate axes that describe a reduced multidimensional space. This becomes particularly useful for the graphical display of multivariate data. Finally, it is to be noted that all the multivariate techniques hold one element in common, which is that they are all based on linear models. This greatly facilitates the solution of equations that would otherwise become rapidly intractable.

7.1.3. Eigenvalues and Eigenvectors

Before we embark on a discussion of multivariate statistical methods, we need to introduce two technical terms that form the basis of computations for most multivariate statistical techniques. These are *eigenvalues* and *eigenvectors*. These terms are encountered with discriminant functions (Section 7.2.1), the method of canonical variate analysis (Section 7.2.3), with factor analysis (Section 7.3.1) and with factor analytic procedures such as principal components analysis (Section 7.3.2).

Eigenvalues and eigenvectors represent two of the more difficult aspects of matrix algebra. This has less to do with the actual computations involved than with an intuitive understanding of the terms. Without unnecessarily belaboring the computational details (since they are readily available elsewhere), the procedures involved are briefly outlined here. As indicated earlier, multivariate techniques generally start with a matrix composed of n-rows and p-columns, which represent simultaneous linear equations, set up as a square[60] matrix X composed of unknown values:

$$a_{11}x_1 + a_{12}x_2 + a_{13}x_3 + \cdots + a_{1n}x_n = \lambda x_1$$
$$\vdots$$
$$a_{p1}x_1 + a_{p2}x_2 + a_{p3}x_3 + \cdots + a_{pn}x_n = \lambda x_n, \qquad [7\text{-}1]$$

which can be set up in matrix form as:

$$AX = \lambda X, \qquad [7\text{-}2]$$

It can be shown that this equation only holds for particular values of λ (Manly, 1994). Setting equation [7-1] equal to zero ($AX - \lambda X = 0$) yields $(A - \lambda I)X = 0$. It can be shown that the determinant of the matrix of coefficients $(A - \lambda I)$ must be equal zero (Van de Geer, 1971). Thus, the values of λ of matrix A can be solved from:

$$\det(A - \lambda I) = 0, \qquad [7\text{-}3]$$

where I is an identity[61] or unit matrix multiplied by λ and det is the determinant. Here the solution of the *characteristic function,* equation [7-3], that is, λ of A, is termed an eigenvalue or *latent root.*[62] There is usually more than one root (λ_1, λ_2, ...λ_n). Once these eigenvalues have been found, they can be substituted back into equation [7-1] to solve for eigenvectors. This procedure entails the use of matrix inversion to produce the required vector of coefficients. This vector is called an eigenvector (other equivalent terms are characteristic, latent or principal vector). This procedure is carried out separately for each eigenvalue. In other words, for each λ of A, there is a column matrix X called an eigenvector (see Manly, 1994; Van de Geer, 1971 for details).

---

[60] A square matrix is one that contains an equal number of rows and columns.
[61] An identity matrix is a square matrix with ones in the main diagonal and zeroes elsewhere.
[62] The term latent root is often associated with factor analysis (Section 7.3.1).

Finally, also important here, is the notion of a symmetric matrix.[63] A unique property of the eigenvectors generated from a symmetric matrix is that they are orthogonal, that is, at right angles to each other when viewed in geometric terms. See Davis, (1986:126-139) for a particularly clear exposition of the details and an elegant if simple geometric interpretation.

7.2. DIFFERENCES BETWEEN GROUPS

Four commonly used methods dealing with group differences are briefly outlined here. They are: [1] discriminant functions, initially developed by Fisher (Fisher, 1936; Lubischew, 1962; Cacoullos, 1973), [2] the generalized distance, or D^2, (Mahalanobis, 1936), [3] canonical variates used in conjunction with discriminant functions (Albrecht, 1980; 1992) and [4] cluster analysis (Parks, 1966; Anderberg, 1973).

7.2.1. Discriminant Functions

Discriminant functions can assist in the placement of unknown specimens into known groups. This is facilitated by the ability to increase the 'discrimination' between groups based on a set of commonly held measurements. However, discriminant functions contain the constraint that group identity and membership within the group must be known in advance, otherwise the use of cluster analysis is indicated (Section 7.2.4). Discrimination is achieved by finding a transformation that maximizes the between-group variation while minimizing the within-variation. This is graphically illustrated in Davis (1986:480) by projecting the samples onto the discriminant function 'line'. The greater the discrimination, the more accurately one can place unknown specimens into the proper group. As the overlap of closely similar groups increases, discrimination is degraded. A consequence is that the percentage of misclassification of individual cases belonging to each group also increases. Criteria that must be met in the development of discriminant functions are that: [1] the membership of each individual within a group must be known and [2] simultaneous membership in more than one group is not allowed. A multivariate normal distribution of all variables is assumed, but difficult to obtain in practice. Nevertheless, discriminant functions have been found to be reasonably robust (Klecka, 1980). Fisher chose coefficients that maximized the 'distance of how far apart' the groups were. That is, these 'discriminants' were computed to maximize the between-group variance relative to the within-group variance. A discriminant function is a linear equation, one per individual specimen, derived from the set of original variables, $x_1, x_2, x_3, \cdots, x_n$, each of which is multiplied by a 'weighing' coefficient, $a_1, a_2, a_3, \cdots, a_n$, such that:

$$F_1 = a_1 x_1 + a_2 x_2 + a_3 x_3 + \cdots + a_n x_n, \qquad [7\text{-}4]$$

where F_1 is the value or score.

---

[63] A symmetric matrix has elements, which are symmetric about the main diagonal. More technically, a matrix is symmetric if it is *square* and equal to its *transpose* (see footnote 65).

A set of these discriminant functions, $F_1, F_2, F_3, \cdots, F_n$, is then computed for each individual case. Each discriminant function score is orthogonal with respect to all others.

7.2.2. Mahalanobis D^2 Statistic

The distance squared between groups (or individual specimens), is termed the 'generalized distance' or D^2 statistic of Mahalanobis (1936), who developed it. The D^2 statistic[64] is of central importance to morphometrics because it describes the *relatedness* or *similarity* between forms, based on multiple *uncorrelated* measurements. (Rao, 1948; see Talbot and Mulhall, 1962 and Lubischew, 1962 for early applications of the D^2). Both the discriminant functions and the D^2 statistic are provided as output in most computer routines. The D^2 statistic is a modified Euclidean distance measure that eliminated the difficulties present with the CRL of Pearson (Section 4.3.1), in that the original measurements are now transformed so that the correlation between variables is reduced to zero (property of orthogonality). If the Euclidean distance between points $\mathbf{X}_i$ ($x_1, x_2, x_3, \cdots, x_i$) and $\mathbf{X}_j$ ($x_1, x_2, x_3, \cdots, x_j$) is d_{ij}, then the squared distance in vector notation is computed from:

$$d_{ij}^2 = \left(\mathbf{X}_i - \mathbf{X}_j\right)'\left(\mathbf{X}_i - \mathbf{X}_j\right),$$

[7-5]

where $\left(\mathbf{X}_i - \mathbf{X}_j\right)$ is the difference vector and equation [7-5] represents the sum of the squared elements of this vector.[65] This equation is simply the generalization of the Pythagorean theorem.[66] Note that it is *uncorrected* for correlations between variables. The D^2 distance, on the other hand takes correlation into account and is set up as:

$$D_{ij}^2 = \left(\mathbf{X}_i - \mathbf{X}_j\right)'\left(\text{cov}_{ij}\right)^{-1}\left(\mathbf{X}_i - \mathbf{X}_j\right),$$

[7-6]

where the (cov_{ij}) is the covariance between the ith and jth variable.[67] This 'variance-covariance' matrix can also be thought of as:

> ... describing the extent to which the Riemmannian hyperspace has to be distorted in order to accommodate the interrelationships existing between the characters when measured in Euclidean space (Blackith and Reyment, 1971:11).

The greater the magnitudes of D^2, the more diverse are the group centroids from each other. Computational details can be found in numerous references (Fisher, 1936;

---

[64] The D^2 statistic can be viewed as the shortest distance (geodesic) between two populations in curved space.
[65] The prime (') refers to the transpose of (x_i-x_j). The transpose in this case, converts a column vector into a row vector in order to facilitate matrix multiplication.
[66] That is, the hypotenuse as the distance, d_{ij}, between points $\mathbf{X}_i$ and $\mathbf{X}_j$.
[67] This covariance term is a metric tensor that represents the inverse of the variance-covariance matrix (matrix of dispersions), which removes the intercorrelations present between the original variables by imposing orthogonality. In effect a distortion of the Euclidean hyperspace.

Blackith and Reyment, 1971; Mardia, *et al.*, 1979; Klecka, 1980; Reyment *et al.*, 1984; Afifi and Clark, 1984).

7.2.3. Canonical Variate Analysis

What remains to be discussed is an allied multivariate procedure, canonical correlation analysis, from which are derived two calculations: [1] the correlation between two *derived*[68] variables and [2] a set of *canonical variates*, or scores, based on the correlation matrix (Van de Geer, 1971). It is these canonical variates or canonical functions (as sums of weighted variables) that are of primary interest here, as they allow for a visual assessment of the relationship between groups. That is, how can the D^2 'distance', as a measure of the differences between groups, be best viewed? This answer is obtained with the help of canonical axes. These canonical axes are derived from a canonical variate analysis, or simply a canonical analysis.[69] While independently computed, they are often a part of discriminant functions software.

Canonical variate analysis has been perceived as a distinct procedure from discriminant functions. Although displayed results are indistinguishable in many respects, the purposes are different (Albrecht, 1992). Discriminant functions are used to place individuals into known groups, while canonical variate analysis has as its primary goal the graphical ordination or visualization of individuals and groups in multidimensional space (Campbell and Atchley, 1981; Blackith and Reyment, 1971; Reyment, *et al.*, 1984; and elsewhere). Both were derived from Fisher (1936) in the sense of *maximizing* the between-group differences, initially between two populations, later more than two, using multiple measurements (Albrecht, 1980).

The method of canonical variate analysis consists of a number of steps. The first step is a rotation of the original axis system. That is, a rigid rotation of the axis system in such a way as to maximize the within-group variation. The observed data points (x, y) as well as group centroids remain unchanged. The second step involves a transformation or rescaling. Here the multidimensional data space is 'deformed' until the within-group variation is made circular. This requires a stretching or compression along each of the axes, resulting in an orthogonal axis system. The third step consists of another rotation, this time to maximize the between-groups variation. The rotated and re-scaled axes are now re-labeled as canonical axes. These canonical axes are orthogonal to each other and can be displayed as 2-D plots, *i.e.*, the horizontal or x-axis as CA-1 and the vertical or y-axis as CA-2. Higher canonical axes can also be visualized as pairs in 2-space. Canonical variables or *roots* can now be extracted (computed) and displayed for each specimen. Specimen groups can be visualized as clouds of data points embedded in this 'deformed' but reduced multidimensional space. These steps usually provide a clearer view of group relationships. Each of these canonical axes contains information derived from the original set of multiple variables. See the papers by Albrecht for useful graphical illustrations of canonical variate analysis (Albrecht, 1980; 1992).

---

[68] A *derived* variable refers to one that represents a weighted combination of other variables (Kachigan, 1991).

[69] For purposes here, one can equate Fisher's discriminant functions (equation 7.4) with canonical variates (functions). See Mardia (1979:319). For this reason some investigators have used the terms *canonical discriminant functions* (Klecka, 1980; Manly, 1994). From these canonical functions, one can extract roots that can then be plotted along orthogonal axes (called canonical axes).

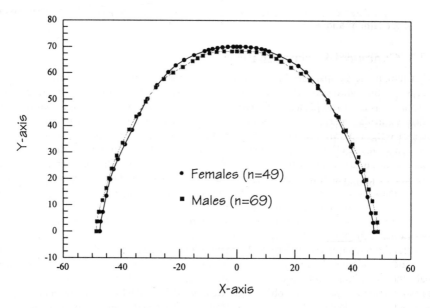

Figure 7.1. Mandibular superimpositions of normal human dentitions.

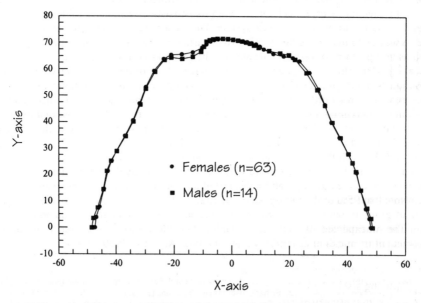

Figure 7.2. Mandibular superimpositions of crowded human dentitions

The example above (Figs. 7.1 and 7.2) illustrates the utility of canonical roots for illustrating the results of discriminant functions. A study of the form of the human mandibular arch using elliptical Fourier functions (Chapter 9) disclosed that not only were there statistically significant differences between crowded and normal dentitions, but also that males differed significantly in shape from females (Ahn, *et al.*, 1999; Lestrel, *et al.*, 1999; Takahashi, *et al.*, 1999). Figures 7.1 and 7.2 depict these form differences. Figure 7.1 illustrates the superimposition of the mean outlines of the normal human dentitions. Figure 7.2 displays the superimposition of the mean outlines of the crowded human dentitions. The data for both the normal and crowded dentitions was normalized for size such that the area under the curve was identical between comparisons (Section 9.3.4). The arches were also mirror-imaged about a horizontal axis to generate a bounded outline (Section 9.3.4).

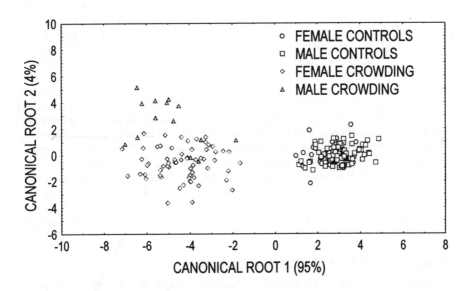

Figure 7.3. A canonical plot of the mandibular data.
Note that 95% of the explained variability resides along the 1st canonical axis

The unexpected finding mentioned earlier, of statistically significant sexual dimorphism in mandibular shape, was expressed with the control females displaying a narrower and longer dental arcade compared to males (Fig. 7.1). However, the pattern was somewhat obscured in the crowded dentition cases (Fig. 7.2).

Utilizing a set 27 amplitude variables, a discriminant function analysis was run, the results of which are shown in Figure 7.3. As mentioned, the data was standardized for size so that only shape differences were involved. Thus, in this case size does not play a significant role with the first canonical axis (along the x-axis) as often assumed with principal components (Section 7.3.2). The cases with normal dentition were clearly separated from those that exhibited crowding. Not unexpectedly, the cases displaying crowded dentitions also exhibited more variability. Finally, the large overlap seen between the female and male controls is also mirrored in the original data (Fig. 7.1).

7.2.4. Cluster Analysis

The next method deals with a more basic problem, the identification of group structure. That is, given a collection of objects are there recognizable sub-groups? Recall that with the use of discriminant functions, group identity must be known in advance.

The purpose of cluster analysis is to find 'clusters' or sub-groups that are presumed to exist within a data set, but have heretofore not been recognized. Cluster analysis removes the constraint that has to be met with discriminant functions and canonical analysis; namely, *a-priori* group identification. Group identification or classification, is one of the main purposes of clustering procedures. Presumably, if the differences within an assemblage of forms are distinct enough, sub groups will appear to be recognized as such. Again, the 'closeness' of the groups to each other indicates distinctiveness or similarity (here as with discriminant functions, the D^2 is a useful measure of this relatedness; although the most commonly used measure is simply the Euclidean distance). A visual device to show these relationships is the dendogram or tree (Parks, 1966). It should be noted that there are numerous different types of 'clustering' algorithms (Anderberg, 1973; Gordon, 1981; Afifi and Clark, 1984; Aldenderfer and Blashfield, 1984). A caution must be inserted here: Different clustering methods may generate different solutions with the same data set. This is partially an outcome of the diverse disciplines, which have contributed to cluster analysis. Thus, it is reasonable to apply more than one clustering method and compare the results. A number of factors can contribute to the performance of various clustering methods: [1] the variables used to describe the data set, [2] the presence of outliers, [3] degree of overlap or relatedness of clusters to each other and [4] the choice of similarity measure such as the Euclidean distance, the D^2, or Pearson's correlation coefficient, as well as other measures (Aldenderfer and Blashfield, 1984).

7.3. DIFFERENCES WITHIN GROUPS

Rather than focusing on differences between groups, as discriminant functions, canonical analysis, generalized distances and cluster analyses do, factor analytic methods and principal components analysis in particular, deal with the *interrelationships* between the measurements themselves. That is, these approaches attempt to identify how those structural dependencies among the variables act to uniquely characterize a group (Comrey, 1973).

7.3.1. Factor analysis

Factor analysis is a technique that has its roots in the social sciences with such disciplines as psychology and education (Cattell, 1965a; 1965b; Kim and Mueller, 1978a; 1978b). Factor analytic methods were initially designed to help social scientists gain a better understanding of the often-complex interrelationships among, what were poorly defined variables. Again, as with clustering approaches, there are various types of factor analytic techniques, and which one is appropriate for a particular data set requires thoughtful consideration.

An investigator may have a number of reasons for undertaking a factor analysis. Such goals may consist of: [1] gaining an understanding of the underlying structure of the variables used; [2] needing to reduce the observed measurements to a smaller, more manageable number, with a minimum loss of information, and [3] identifying what this set of smaller, uncorrelated, variables (factors) explains in terms of variability.

To plan and carry out a factor analytic procedure, the following steps are usually followed: [a] initial selection of variables, [b] computation of a correlation matrix, [c] extracting unrotated factors, [d] generating a rotated solution and [e] interpreting this rotated solution (Comrey, 1973).

In parallel with other multivariate methods, factor analysis is based on a linear model. In this case, one starts with a set of standardized variables. That is, each original variable, $X_1, X_2, X_3, \cdots, X_p$, is standardized with a mean of zero and a variance of one (Afifi and Clark, 1984). These standardized values are now called 'standard scores'. Each original X_p is rewritten as:

$$X_p = L_{p_1} F_1 + L_{p_2} F_2 + L_{p_3} F_3 + \cdots + L_{p_n} F_n + e_p,$$ [7-7]

where the $F_1, F_2, F_3, \cdots, F_n$ values represent the common factor scores, the L_{ij} coefficients, called the factor loadings, are weighing terms, and the $e_1, e_2, e_3, \cdots, e_p$ are the unique factors or error terms associated with each original variable. It should be mentioned that the number of factor scores are usually considerably less in number than the original variables. Only the first few factors are usually considered.

Since the variance of all of the components is equal to one, it is possible to partition the factor model into two parts: [1] the communality, h^2, which is the common factor variance, and [2] the specificity, which is the variance due to the unique factor or error term. From a computational viewpoint, one is interested in the factor loadings, L_{ij}, and in the communalities, h^2, as these are central to the factor extraction process (Afifi and Clark, 1984). The communalities provide a method by which one can assess the percentage of the variance explained by each of the original variables. The factor loadings are actually the correlations between the standardized variables and the factors. At this point, it is helpful to visually display the results in a factor diagram. This is simply a Cartesian scatter plot where instead of the original variables; their factor loadings (composed of a weighing of the original variables) are plotted against the respective factors (common factors) where the factors are orthogonal, or at 90 degrees, to each other. This is commonly displayed with the horizontal or x-axis representing factor 1, and the vertical or y-axis as factor 2. In this way, one can see the 'structure' in the original variable set; that is, how they 'load' onto the factors. At this point one can also 'rotate the factors'. This is often done in an attempt to make the loadings more interpretable. A number of rotations are in common use, such as the varimax procedure or the oblique rotation, both found in most computer programs.

7.3.2. Principal Components Analysis

We now turn to an allied method called principal components analysis (PCA), which is frequently used in morphometrics. The difference between these two computationally similar procedures, factor analysis and PCA, has engendered considerable confusion. The major emphasis of factor analysis is to obtain a small number of easily interpretable variables, or factors, that reveal the essential information (underlying structure) present in the data set. These factors are intended to explain the interrelationships between the original variables. In contrast, PCA is primarily focused on extracting a set of components that explain as much of the total variance as possible.

From a morphometric perspective, one practical reason for using principal components is that as the number of principal components increase, the information being explained, in terms of variability, rapidly decreases. Generally, only a comparatively small number of the uncorrelated components are needed to explain a majority of the variance present. Thus, if the groups can be discriminated reasonably well from each other, then they should also plot out as discrete clouds of points (Blackith and Reyment, 1971; Reyment, *et al.*, 1984).

As in factor analysis, a set of uncorrelated variables is generated. A PCA consists of a rotation of the original variable set's axis system, to a new set of orthogonal axes, which are called principal axes, where these new axes coincide with the maximum variation in the original data set. For each specimen a 'principal component' is computed as a linear combination of the original variables. These original values are first subtracted from the sample means resulting in means equal to zero which is analogous to factor analysis; however, the sample variances are not altered in contrast to factor analysis where they were standardized to one. As the total variance accounted for remains the same after transformation, one can arrange the principal components in decreasing order based on the variance (Afifi and Clark, 1984). As these principal components are orthogonal, or independent of each other, this allows (as in factor analysis) the graphing of each principal component score, one for each specimen in the sample. Thus, one can plot the first principal component along the x-axis and the second principal component as the y-axis. Higher principal components are displayed in a similar pairwise manner. This approach is analogous to the visualization of canonical axes used with discriminant functions (Section 7.2.1.).

Experimental work on growth studies has suggested that the first principal component is often a measure of total organism size. Recalling that the first principal component explains the largest percentage of the variance, this would indicate that size has an overwhelming effect. This has led to a number of investigators to state that principal components will allow the separation of size from shape (Jolicoeur and Mosimann, 1960; Blackith and Reyment, 1971; Reyment, *et al.*, 1984; Spivey, 1988). That is, 'size' is presumed to load along the first principal axis and 'shape' is distributed along the higher numbered axes. Nevertheless, this may be an oversimplification and not necessarily always the case (Reyment, 1985; Lestrel, *et al.*, 1989; Read, 1990). Moreover, there is the issue of what is meant by 'shape' with the use of conventional metrics (distances and angles). Such measurements may not be reflective of 'shape' as commonly presumed. This issue is taken up in the next two chapters.

7.4. SOME FINAL COMMENTS

The six techniques briefly outlined in Sections 7.2 and 7.3, are independent of the type of data sets and variables chosen. Nevertheless, while most multivariate morphometric studies to date tend to be based on data sets composed of distances and angles (Section 8.2), they are not limited to those metrics.

An initial study was based on lateral x-ray images of the cranial base in four groups[70] of primates. Using Fourier amplitudes (Section 9.3.3) produced, in spite of considerable variability,[71] an almost perfect classification with discriminant functions (Lestrel, *et al.*, 1988). Later, baboons were added raising the number of groups to five (Lestrel and Swindler, 1996). These two studies are described in more detail in Section 9.5.6. Another Fourier analytic study (Section 9.5.6) utilizing discriminant functions, demonstrated that using variables derived from the cranial base, maxilla, and mandible, produced a 100% correct classification (Lestrel, *et al.*, 1996a; Lestrel, *et al.*, 1996b; Surillo, 1996; Mizraji, 1996). However, if only one or two of the craniofacial structures were used, instead of all three, then the misclassification rate increased. Finally, an earlier study of the shape of the distal end of the primate femur using Fourier descriptors, displayed functional-behavioral characteristics that could be discerned from the canonical plots (once size had been controlled for), functional aspects that would have been considerably more difficult to detect with conventional metrics because of the difficulty of removing the confounding effect of size (Lestrel, *et al.*, 1977). These approaches, discussed here, of using Fourier descriptors to deal with shape analysis, lead to boundary morphometrics, which are discussed in Chapter 9. The next chapter focuses on landmark-based representations, which are subsumed under coordinate morphometrics.

KEY POINTS OF THE CHAPTER

This chapter dealt with the initial approaches used to analyze morphometric data; namely, the application of multivariate statistics to data sets encountered in the biological sciences. Two major directions can be discerned, statistical between-group and within-group considerations. These in turn have spawned a number of multivariate techniques, some of which were discussed in this chapter.

CHECK YOUR UNDERSTANDING

1. Why is an analysis based on multiple variables better than one based on a single variable? Can you think of cases where one variable could be satisfactory?

2. When dealing with multiple variables, what does the presence of correlations imply? What can be done to reduce the effects of correlated variables in an analysis?

---

[70] Specimens included chimpanzees, gorillas, macaques and humans.
[71] This variability was a consequence of the presence of infants, juveniles as well as adults in the data.

3. What is the major difference between discriminant functions and cluster analysis? Does this difference determine which method should be used?

4. What are the differences between factor analysis and PCA?

5. Work through pages 126-139 of Davis (1986) to gain a clearer understanding of the meanings of eigenvalues and eigenvectors. How are eigenvalues used in factor analysis and discriminant function analysis?

6. Utilizing a statistical package and sample data of your own choosing, run discriminant functions and make a canonical plot of variables one and two. Then using the same data set run a factor analysis. Compare the output from the two runs. What conclusions can you draw?

7. Utilizing your sample data, run the cluster programs contained in a statistical package. Are there significant differences between the different clustering techniques available? Can these be reconciled?

8. What are the differences between the Euclidean distance measure and Mahalanobis' D^2 statistic? Do a literature search to identify some of the other distance measures that also have also been used.

REFERENCES CITED

Afifi, A. A. and Clark, V. (1984) *Computer-Aided Multivariate Analysis*. New York: Van Nostrand Reinhold.

Ahn, S. S., Lestrel, P. E. and Takahashi, O. (1999) The presence of sexual dimorphism in the human dental arches: A Fourier analytic study. *J. Dent. Res.* **78**:278.

Albrecht, G. H. (1980) Multivariate analysis and the study of form, with special reference to canonical variate analysis. *Am. Zool.* **20**:679-693.

Albrecht, G. H. (1992) Assessing the affinities of fossils using canonical variates and generalized distances. *Hum. Evol.* **7**:49-69.

Aldenderfer, M. S. and Blashfield, R. K. (1984) *Cluster Analysis*. Quant Appl. Soc. Sci. Ser. No. 44. Sage Pub. Inc. Beverly Hills.

Anderberg, M. R. (1973) *Cluster Analysis for Applications*. New York: Academic Press.

Bernstein, I. H. (1988) *Applied Multivariate Analysis*. New York: Springer-Verlag.

Blackith, R. E. and Reyment, R. A. (1971) *Multivariate Morphometrics*. New York: Academic Press.

Cacoullos T: (1973) *Discriminant Analysis and Applications*. New York: Academic Press.

Campbell, N. A. and Atchley, W. R. (1981) The geometry of canonical variate analysis. *Syst. Zool.* **30**:268-280.

Cattell, R.B. (1965a) Factor analysis: an introduction to essentials. I. The purpose and underlying models. *Biometrics* **21**:190-215.

Cattell, R.B. (1965b) Factor analysis: an introduction to essentials. II. The role of factor analysis in research. *Biometrics* **21**:405-435.

Comrey, A. L. (1973) *A First Course in Factor Analysis.* New York: Academic Press.

Davis, J. C. (1986) *Statistics and Data Analysis in Geology* (2nd Ed). New York: John Wiley.

Fisher, R. A. (1936) The use of multiple measurements in taxonomic problems. *Ann. Eugen. (Lond.)* 7:179-188.

Gordon, A. D. *Classification.* (1981) London: Chapman and Hall Ltd.

Hotelling, H. O. (1931) The generalization of Student's ratio. *Ann. Math. Statist.* 2:360-378.

Hotelling, H. O. (1933) Analysis of a complex statistical variables into principal components. *J. Edu. Psych.* 24:417-441 and 498-520.

Jolicoeur, P. and Mosimann, J. E. (1960) Size and shape variation in painted turtle: A principal components analysis. *Growth* 24:339-354.

Kachigan, S. (1991) *Multivariate Statistical Analysis.* (2nd Ed.) New York: Radius Press.

Kim, J. and Mueller, C. W. (1978a) *Introduction to factor analysis.* Quant Appl Soc Sci. Ser. No. 13. Sage Pub. Inc. Beverly Hills.

Kim, J. and Mueller, C. W. (1978b) *Factor analysis.* Quant Appl Soc Sci. Ser. No. 14. Sage Pub. Inc. Beverly Hills.

Klecka, W. R. (1980) *Discriminant analysis.* Quant. Appl. Soc Sci. Ser. No. 19. Sage Pub. Inc. Beverly Hills.

Lestrel, P. E., Kimbel, W. H., Prior, F. W. and Fleischmann, M. L. (1977) Size and shape of the hominoid distal femur: Fourier analysis. *Am. J. Phys. Anthrop.* 46:281-290.

Lestrel, P. E., Stevenson, R. G. and Swindler, D. R. (1988) A comparative study of the primate cranial base: elliptical Fourier functions. *Am. J. Phys. Anthrop.* 75:239.

Lestrel, P. E., Sarnat, B. G. and McNabb, E. G. (1989) Carapace growth of the turtle *Chrysemys scripta:* a longitudinal study of shape using Fourier analysis. *Anat. Anz. (Jena)* 168:135-143.

Lestrel, P. E., Mizraji, G. M. and Surillo, S. A. (1996a) A classification of skeletal jaw relationships and considerations of variability: Fourier descriptors. *J. Dent. Res.* 75:337.

Lestrel, P. E., Surillo, S. A. and Mizraji, G. M. (1996b) A Re-evaluation of Sassouni's classification of skeletal jaw relationships using elliptic Fourier functions. *J. Dent. Res.* 75:337.

Lestrel, P. E. and Swindler, D. R. (1996) The numerical characterization of the primate cranial base: A comparative study using Fourier Descriptors. *Am. J. Phys. Anthrop.* Suppl. 22:148.

Lestrel, P. E., Takahashi, O. and Ahn, S. S. (1999) An Analysis of dental arch form: A model based on Fourier descriptors. *J. Dent. Res.* 78:278.

Lubischew, A. A. (1962) On the use of discriminant functions in taxonomy. *Biometrics* 18:455-477.

Mahalanobis, P. C. (1936) On the generalized distance in statistics. *Proc. Nat. Inst. Sci. (India)* **2**:49-55.

Manly, B. F. J. (1994) *Multivariate Statistical Methods - A Primer.* (2nd Ed.). London: Chapman and Hall.

Mardia, K.V., Kent, J. T. and Bibby, J. M. (1979) *Multivariate Analysis.* London: Academic Press.

Mizraji, G. M. (1996) *Skeletal jaw variability: Elliptical Fourier descriptors.* Unpublished MS thesis, Oral Biology, UCLA School of Dentistry.

Mosimann, J. E. (1970) Size allometry: size and shape variables with characterizations of the lognormal and generalized gamma distributions. *J. Am. Stat. Assoc.* **65**:930-945.

Nadler, M. and Smith, E. P. (1993) *Pattern Recognition Engineering.* New York: John Wiley & Sons.

Parks, J. M. (1966) Cluster analysis applied to multivariate geologic problems. *J. Geol.* **74**:703-715.

Pearson, K. (1901) On lines and planes of closest fit to a system of points in space. *Phil. Mag.* **2**:557-572.

Pearson, K. (1926) On the coefficient of racial likeness. *Biometrika* **18**:105-117.

Penrose, L. S. (1954) Distance, size and shape. *Ann. Eugen.* **18**:337-343.

Porter, T. M. (1986) *The Rise of Statistical Thinking 1820-1900.* New Jersey: Princeton University Press.

Read, D. W. (1990) From multivariate to qualitative measurement: Representation of shape. *Hum. Evol.* **5**:417-429.

Reyment, R. A. (1985) Multivariate morphometrics and analysis of shape. *Math. Geol.* **17**:591-609.

Reyment, R. A. (1991) *Multidimensional Paleobiology.* Oxford: Pergamon Press.

Reyment, R. A., Blackith, R. E. and Campbell, N. A. (1984) *Multivariate Morphometrics* (2nd Ed.). New York: Academic Press.

Sneath, P. H. A. and Sokal, R. R. (1973) *Numerical Taxonomy.* San Francisco: W. H. Freeman and Co.

Spearman, C. (1904) 'General intelligence' objectively determined and measured. *Am. J. Psych.* **15**:201-293.

Spivey, H. R. (1988) Shell morphology in barnacles: quantification of shape and shape change in *Balanus. J. Zool. (Lond.)* **216**:265-294.

Spoehr, K. T. and Lehmkuhle, S.W. (1982) *Visual Information Processing.* San Francisco: W. H. Freeman.

Sprent, P. (1972) The mathematics of size and shape. *Biometrics* **28**:23-28.

Surillo, S. A. (1996) *A quantitative assessment of size and shape in skeletal jaw classifications using elliptical Fourier functions.* Unpublished MS thesis, Oral Biology, UCLA School of Dentistry.

Takahashi, O., Lestrel, P. E. and Ahn, S. S. (1999) A size and shape study of dental arch crowding versus non-crowding: Fourier analysis. *J. Dent. Res.* **78**:278.

Talbot, P. A. and Mulhall, H. (1962) *The Physical Anthropology of Southern Nigeria.* Cambridge: Cambridge Univ Press.

Van de Geer, J. P. (1971) *Introduction to Multivariate Analysis for the Social Sciences.* San Francisco: W. H. Freeman and Co.

8. COORDINATE MORPHOMETRICS

The Latest authors, like the most ancient, strove to subordinate the phenomena of nature to the laws of mathematics.

Sir Isaac Newton (1642-1727)

8.1. INTRODUCTION

Coordinate morphometric methods share the characteristic of being based solely on data points composed of 2-D (x, y) or 3-D (x, y, z) coordinates, usually in the Cartesian system. Relationships *between* data points, when considered, are generally viewed as connected with straight lines that make no attempt to follow the actual boundary outline. Thus, these methods ignore the boundary of the form. Boundary considerations are taken up in Chapter 9. Five landmark-based representations will be examined. These are: [1] the conventional metrical approach consisting of distances, angles and ratios (CMA); [2] biorthogonal grids (BOG); [3] finite elements (FEM); [4] thin-plate splines (TPS) and [5] Euclidean distance matrix analysis (EDMA). The latter four were developed within the last two decades as answers to some of the deficiencies posed by conventional metrics. The conventional method (CMA) has been the predominant approach and continues to be universally used. Biorthogonal grids and finite elements have had increasing adherents. Thin plate splines represent a newer approach that, like its precursor, biorthogonal grids, attempts to place D'Arcy Thompson's *transformation grids* on a more solid mathematical footing (Thompson, 1915; 1942; Bookstein, 1977; 1997). Finally, it must be noted that biorthogonal grids, finite elements, and thin-plate splines, are methodologies that focus on changes and model the form as a *deformation*. This will be made clearer when these methods are examined in more detail.

8.2. THE CONVENTIONAL METRICAL APPROACH

The first representation to be considered is the universally used metrical approach (CMA), consisting of distances, angles and ratios. The raw data of CMA are points or *landmarks,* which are presumed, if not always clearly recognized as such, as satisfying the criterion of homology. That is, these points must be, in some sense, 'homologous points'.

8.2.1. The Issue of Point Homology

Homology in geometric terms has been defined as a one-to-one mapping between corresponding points on different forms (Bookstein, 1991). Moreover, such points must be *biologically equivalent* between organisms and samples of organisms; otherwise, such point relationships simply become a geometric exercise, interesting in formal terms, but largely biologically irrelevant.

173

Traditionally, homology was never considered in terms of landmarks or points but rather morphological structures. It was first formalized by Owen in 1843:

> ... as the same organ in different animals ... (Quoted in Kaesler, 1967).

As Owen's definition lacked a phylogenetic outlook, it needed correction. After Darwin it was modified and became:

> ... homology is resemblance due to inheritance from a common ancestor (Simpson, 1961:78).

Thus, in biological terms, homology has to be based on ontogeny or phylogenetic considerations (Cartmill, 1994). Considerable controversy still surrounds these two contrasting definitions of homology; that is, one geometric and the other biologic (O'Higgins and Johnson, 1988). That is, can homologous points even be related to changes in homologous structures? Comparison of forms, based only on a few homologous landmarks, can lead to the misleading conclusion that little in the way of changes in size and shape are present (due to growth, phylogeny, teratology, etc.) because the two landmarks are only slowly changing, in contrast to rapid change of the form *between* the landmarks. Consider a structure lying between two points that 'balloons out' during growth. Alternatively, the question arises whether relative shifts in landmark position between forms are subject to the processes of growth, evolution or other factors, or are they just being carried along by alterations in the broader aspects of the form itself? For example, within the craniofacial complex, there are certain landmark locations that are largely independent of the curvature in which they are embedded. Consider the variability in the position of *bregma* in relationship to the parietal bones of the cranial vault as viewed in the sagittal aspect (see Fig. 1 in O'Higgins and Johnson, 1988). This suggests that major changes in homologous points could occur between forms, while those homologous structures (in which the points are embedded) may be more or less stable and unchanging for long periods of time. Given the complex ways in which the shapes of organisms are altered with growth and other forces, limiting one's analysis to homologous landmarks may not be the most optimal way of extracting the biologically relevant information that resides in the form.

Three types of homologous landmarks have been proposed (Bookstein, 1991): [1] juxtaposition of tissues, [2] maxima of curvature and [3] extremal points. Juxtaposition of tissues apparently is intended to refer to the boundary between distinct regions where sharp discontinuities occur. Maxima of curvature would include well-characterized features of the form such as the tip of the nose, tips of incisors, etc. Extremal points refer to a whole class of points that arise in morphometrics. Examples are end points of diameters or maximum length, and many of the usual cephalometric landmarks such as *menton, pogonion, articulare*, etc. According to Bookstein, many of these points in the third category are not 'true' landmarks because of difficulties with homology between specimens and the lack of biological explanation (Bookstein, 1991).

Problems may also arise with respect to the variability of the above landmark types, in particular, their presence or absence. The latter point may especially be a problem when a growth series is examined. Besides 'homologous' landmarks mentioned above, another class of points can be identified on outlines (White and Phenice, 1988). These are often called pseudo-landmarks, in the sense that they are determined from geometrical rules (bisections, trisections and the like) using other pseudo-landmarks as well as homologous landmarks.[72] Such points play an important role with boundary outline methods (Chapter 9).

8.2.2. The Conventional Method (CMA)

The absence of approaches that can precisely measure the shape of irregular forms, such as those present in biological forms, has of necessity, focused on homologous points or landmarks. Thus, reasons for the widely used approach (CMA) are not hard to find. They include factors such as: conveniently located measures, historical tradition and an assumed valid representation of forms. It must be emphasized at the outset that CMA was originally developed for measuring regular geometric (artificial) objects and its widespread application to irregular natural forms may not be the most appropriate, in spite of its ease of use. The defects of each of these measures (linear distances, angles and ratios) are briefly sketched below.

The presence of considerable size differences between forms can also act to confuse the analysis of shape and shape changes, suggesting that ways are needed to minimize or remove the potentially confounding effect of size. Moreover, it can be argued that in many cases it is the biologically-based shape information, in contrast to size, which may be of more interest from a morphometric perspective. Various approaches have been proposed to control for the effect of size. The simplest way of traditionally normalizing the morphological form for 'size' has been to take a single measurement such as height or length as a standard and adjust other specimens by a multiplying factor. While this may seem satisfactory, it must be emphasized that only a one-dimensional organism, in effect a line, can be exactly standardized with a single length. Area is the proper standardization for 2-D figures and volume for 3-D structures (solids). This stricture is often violated or ignored. When area *is* chosen, the use of CMA is often based on the improper assumption that complex irregular forms can be adequately circumscribed using simple geometric figures such as triangles, circles, ellipses or rectangles (Lestrel, 1974; Read, 1990).

8.2.3. Use of Angles and Ratios as Shape Measures

Angles have been often cited as being 'size-independent', which while technically correct, has led to their use as measures of 'shape'. However, this usage is misleading and, in fact, what an angle measures is open to question. The actual shape (or outline) lying *within* an included angle, can be anything, it is simply not being measured.

The following example is intended to illustrate how 'shape' information has been commonly extracted with CMA. The commonly used cranial base angle (*Nasion-Sella-Basion*, or *N-S-BA*), illustrated in Fig. 8.1, to measure the angulation of the cranial base is a case in point. Biological issues aside, such as whether *Nasion* is even a cranial base landmark, there is the very real question of *where* the shape change is occurring. That is, if the *N-S-BA* angle, say, increases, is *Sella* moving inferiorly? Alternatively, is *Basion*

---

[72] See the discussion of pseudo-landmarks in Dryden and Mardia (1998).

or *Nasion* moving superiorly? Without independent confirmation those questions remain unanswerable.

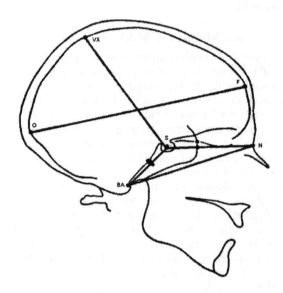

Figure 8.1. The conventional metrical approach.
Symbols: BA, *Basion*; S, *Sella*; N, *Nasion*; VX, *internal vertex*;
O, internal occipital aspect; F, internal frontal aspect, such that
the distance O-F measures the maximum extent.

Ratios, as well as angles, are also problematical as 'shape' measures. The widespread utilization of readily obtained ratios has tended to obscure both theoretical as well as biological difficulties. Ratios, as compound measures of gross proportions, have also been used as 'shape', that is, as scaling variables to control for size differences. Such ratios or indices abound in the biological literature.

A number of theoretical difficulties are attendant with ratios. One is that statistical difficulties arise because the ratio x/y is not a linear function of x and y, unless one takes logarithms (Hills, 1978). Another one is the loss of normality (Atchley, *et al.*, 1976). The use of ratios has sparked considerable debate (Corruccini, 1975; Atchley, *et al.*, 1976; Dodson, 1978). Further, the problematic use of such ratios seems to be insufficiently appreciated in spite of the criticisms that have been leveled by numerous investigators (Albrecht, 1978; Albrecht *et al.*, 1993; Corruccini, 1973; 1977; Atchley, 1978).

Consider the following examples of the human occipital aspect as measured with the two occipital indices (Fig. 8.2, adapted from Tobias, 1959). Tobias compared two indices: [1] Pearson's occipital index and [2] Wagner's ratio. Pearson's index is defined as the ratio of the curvature of the occipital bone from *lambda* to *opisthion* in *Norma lateralis*, to the *lambda-opisthion* cord (first column in Table 8.1). Wagner's ratio is the ratio of the chord to the arc (second column in Table 8.1). Note that these indices are poor shape descriptors and do not even begin to distinguish between children and adults (Lestrel, 1974).

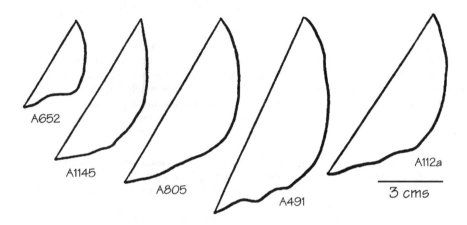

Figure 8.2. Form variation in the occipital aspect.

A652	Nyasa infant	80.3	57.3
A1145	Immature Bushman	80.0	57.1
A805	Xosa child	80.3	57.2
A491	Zulu male adult	80.1	57.2
A112a	Bushman male adult	80.2	57.2

Table 8.1. Measures of occipital form using indices.

Ratios are also tricky from another point of view. Consider the commonly used numerical representation (a rectangle) for a molar, BL x MD (buccolingual width times mesiodistal length), where 'size' is defined as the area of a rectangle and 'shape' is the ratio, or MD/BL. Besides the theoretical difficulties with ratios, mentioned above, there is the issue that molars, as well as other teeth, are not 'rectangles'. The rectangle is only a rough approximation of tooth form, and cannot provide any information about the outline, which may be of considerable ontogenetic and/or phylogenetic importance. In other words, the 'rectangle' representation is an example of an incomplete mapping (Read, 1990). While the rectangle simplification facilitates measurement, it is at a cost of loss of detail in the outline and with it, potentially important shape information.

8.2.4. Some Other Deficiencies with CMA

In the defense of CMA, it can be argued that if the form exhibits complete geometric regularity, then CMA can be profitably applied. By geometric regularity, what is meant is that forms are composed of simple geometries such as circles, triangles, rectangles, etc., and these can be quite complex when combined, consider a geodesic dome.

It is also readily admitted that some elegant and fruitful approaches with CMA have appeared. A case in point is the use of *pattern profile analysis* developed by Garn and colleagues (Garn, *et al.*, 1984). In this procedure a set of ordered measures, in this case cephalometric measurements taken from lateral x-rays of the human head are selected and re-expressed as *z-scores* (standard deviation units). That is, the each measurement is subtracted from the mean for their sex and age group and divided by the appropriate standard deviation. This z-score determines how much, larger or smaller, in standard deviation terms, that measurement is from the mean. These z-scores can than be graphed as an ordered series to depict the closeness or departure from the mean of a morphological structure. Such figures are nicely illustrated in Garn, *et al.* (1984).

However, the application of CMA to natural (in contrast to artificial) forms for describing size and shape remains generally insufficient. This arises because: [1], most biological morphologies are irregular in form, [2] a limited number of isolated landmarks are usually employed resulting in a sparse or inadequate sampling[73], [3] there is an unavoidable bias present due to the subjectivity involved in the choice of landmarks (*i.e.*, often based on convenience rather than chosen to reflect biological variability or other criteria), [4] only a small percentage of the visual information is being sampled, [5] difficulties in adequately standardizing for size and [6] one is unable to subsequently reproduce the morphology from the CMA measurements (information-preserving property).

Moreover, with respect to the issue of representation, it can be demonstrated that the CMA is an incomplete mapping; and consequently, much of the information present in a form, potentially of biological significance, is not being measured. This is an especially acute problem if one is interested *in numerically describing morphological shapes and changes in shape,* and the form only contains a few sparsely located homologous landmarks. Thus, although with CMA some of the visual information in the form can always be quickly and conveniently extracted, it remains necessarily incomplete. In sum, the continuing widespread use of CMA is regrettable in that we seem to have "placed the cart before the horse" so to speak. That is, too little emphasis has been placed on the measurement system *per-se,* in contrast to other perhaps seemingly more legitimate concerns having to do with the issues being investigated. Consequently, there is the distinct possibility that results using CMA may be at best limited and at worst misleading or incorrect. These and other factors stimulated the development of alternative methods to circumvent the deficiencies of CMA.

As indicated earlier, these problems attendant with CMA, stem from a very basic issue. CMA is a representation system that does not correspond (map) precisely with the form being numerically described. That is, the measurements used tend to be arbitrarily

---

[73] See the work of Moss and Young (1960) for an example dealing with the limitations of CMA in the analysis of the curvature of the cranial vault. This is a consequence of the presence of large areas of smooth curvature remaining undefined, as there are no landmarks available (Lestrel, 1974; O'Higgins, 1989).

selected, often too few in number and, therefore, cannot be used to precisely recreate the outline of the form under consideration. Thus, the deficiencies arising with the use of CMA to measure shape and shape changes can be summarized as: an inappropriate mapping between the morphology and its numerical representation and the reliance on a measurement system designed for regular geometric objects and never intended for complex irregular forms of the type encountered in the biological sciences.

8.3. BIORTHOGONAL GRIDS

The biorthogonal grids approach (BOG) was developed by Bookstein to deal with two aspects of morphometrics (Bookstein 1977; 1978; 1982; 1983; 1986; and elsewhere). These were to: [1] allay many of the shortcomings of CMA and [2] provide a more effective numerical solution to the well-known D'Arcy Thompson's transformation grids, which have been viewed as limited because they were largely qualitative in nature (see Medawar, 1945 for an earlier attempt). The BOG technique denotes a homologous-point representation, an aspect, which it shares with CMA. Beyond that, it is an entirely different approach from CMA.

8.3.1. Basis for Biorthogonal Grids

The foundation of the BOG method is a comparison between two 2-D forms, one of which is designated a base form and the other one which reflects shape changes from the base form. These shape changes are viewed as 'deformations'. The technique computes estimates of the *difference* in form between the two objects. Starting with the base form, we can construct a series of triangles across the morphology connecting homologous points (Fig. 8.3). Focusing on just one of these triangles in the base form (called the initial or base triangle), consider a circle circumscribed inside the triangle such that it touches the three sides. Now, if this triangle is deformed (a shape change) in some fashion, then the same circle will now be distorted into an ellipse. The major, *da*, and minor, *di*, axes of this ellipse can be viewed as estimates of this shape change. They represent the 'principal dilations' or estimates of maximum 'stretch' and 'shrinkage' due to this deformation. Figure 8.3 (adapted from Bookstein, 1982) illustrates this deformation of a circle inscribed within a 'reference' triangle.

In Figure 8.3A, this circle is deformed into the ellipse shown in Figure 8.3B. The ratio of the lengths in triangle A'B'C' divided by the corresponding lengths in triangle ABC are called dilations. The dilation of 1.12 in 8.3B represents a 12% 'stretching' along that line, while the dilation of 0.79 represents a 21% 'shrinking' in the orthogonal or perpendicular direction. The ratio of the major to minor axis can be considered as an estimator of the shape change (Reyment, 1991). That is, 'size change' can be computed as the sum of logarithms, *log (da)* + *log (di)*. While the 'shape change' is calculated from the difference between logarithms, *log (da)—log (di)*. The 'cross' denoting the center of the ellipse with its major and minor diameters, represents a tensor. This 2-D tensor is defined as a measure of the deformation arising simultaneously in all directions, of which the principal dilations measure the maximum and minimum values. A requirement of this method is that the deformation is considered *uniform* within the triangle boundary. That is, the principal dilations are based on a constant tensor field. This is the property of homogeneity and forms the basis for the assumption that shape change is uniform (Bookstein, 1984, and elsewhere). By forming ratios of these dilations, the result is invariant with respect to the coordinate system. By generating a series of triangles connecting homologous points across the morphological form, one can compute pairwise comparisons with the base triangle and its deformed pair.

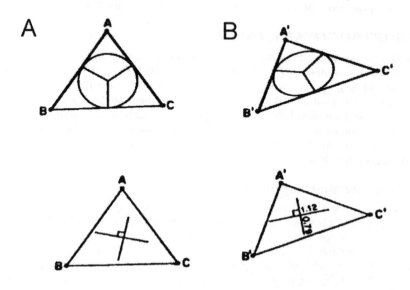

Figure 8.3. Biorthogonal grid triangles.
(See text for explanation)

8.3.2. Constraints of Biorthogonal Grids

Deformation techniques such as BOG focus on 'shape change' in contrast to 'shape' itself. This can be an advantage in that the initial requirement of a numerical description of shape is sidestepped, although at the expense of considerably increased abstraction (*e.g.,* the dilations) which are difficult to relate back to the observed changes in actual shape.

As a general numerical model, BOG remains incomplete. Like CMA, it makes no provision for the curve information that lies between homologous landmarks (Chapter 9). In reference to cephalometrics (using CMA), Bookstein correctly stated that:

> To the extent that cephalometrics is based wholly upon landmark location and quite ignores the curving outline there and in between, it cannot possibly measure individual form in a thorough fashion (Bookstein, 1978:18).

Nevertheless, Bookstein also has chosen to largely dismiss the boundary information from consideration:

> Information about curving form between landmarks, available from tracings of continuous biological outlines, is excluded from consideration here (Bookstein, 1984:480).

And also:

> To pass from the biological to the biometrical context, homology must be considered as a *mapping function*, a correspondence relating *points to points* rather than parts to parts (Bookstein, 1991:57).

It should be reiterated that this definition of point homology is at variance with the traditional biological definition of homology as discussed earlier. Thus, this definition while intended to be rigorous from a mathematical perspective, is one, with which many biologists would quarrel with as being too narrow, with an overemphasis on geometric requirements, at the expense of biological considerations. Even 'simple' morphologies are much too complex to be easily reduced to a small set of homologous landmarks.

A number of other constraints are associated with BOG. If one of the angles in the triangle approaches 0 or 180 degrees, then the BOG results become increasingly unreliable (Bookstein, 1982). Because the sampling of genuinely existing homologous points tends to be sparse, especially with respect to craniofacial structures, the number of triangles that can be constructed tend to be small and cover large aspects of the morphology. To alleviate this problem of too few adequate homologous landmarks, Bookstein has argued for the inclusion of a class of arbitrary 'pseudo-landmarks' between homologous points (see Reyment, 1991:127 for an example using a microfossil foreminifer *Brizalina* specimen). Rigorously defined however, such points are not admissible in this system since they violate the required criterion of a one-to-one mapping between homologous points (Lestrel, 1989b). Nevertheless, homology as discussed at the beginning of this Section, whether point or structure (Cartmill, 1994), remains a vexing and controversial issue.[74] Other difficulties with BOG include the requirement of a homogenous or constant tensor field across the plane of the triangle (*i.e.,* the requirement of zero curvature). Whether this uniformity of shape change, based on this constant tensor field is warranted, has been questioned. There is considerable evidence that the assumption of uniform shape change (homogeneity) is an unlikely situation with real data (Mardia and Dryden, 1989). Finally, although the above material may be viewed as seemingly unduly critical in its assessment of BOG, it should be noted that all methods have limitations and BOG is no exception. In a more positive vein, it can be asserted that BOG represented the first approach to successfully characterize Thompson's transformation grids in much more rigorous terms.

8.4. FINITE ELEMENT ANALYSIS

The finite element method (FEM) was borrowed from engineering where it is a well-established procedure for measuring the effects of loadings on designed structures such as buildings, ships or aircraft. When a structure is loaded, stresses are set up which act to deform the material. This loading results in physical changes or displacements, which are measurable. The magnitude of the deformation is dependent on the material,

---

[74] Reference should be made to Sneath and Sokal (1973) for a discussion of homology.

its density as well as its geometric shape. It is apparent that under loading, a structure, which is uniform, will behave differently from one that has varying density. For example, the cranial bones have been considered in an analogous way to a thin shell structure.

8.4.1. Finite elements in 2-D

The application of FEM here views the deformation that occurs under loading to result in measurable physical changes located within the morphology (Bathe and Wilson, 1976; Huiskes and Chao, 1983). FEM, as used to numerically describe morphological shape changes, is very similar, if not mathematically identical, to BOG. It is also limited to homologous landmarks and is invariant with respect to the coordinate system. It differs in that it can be extended to 3-D. It shares with BOG the concept of deformation. Recalling that BOG defines the shape change as a deformation, this can be viewed as a circle (embedded in a triangle) being distorted into an ellipse. With FEM the major and minor or 'principal dilations' are now 'strain' measures of a rigid body. Figure 8.4 shows the FEM applied to the lateral aspect of the rat. Results of such a rat study are briefly described later (Moss, *et al.*, 1987).

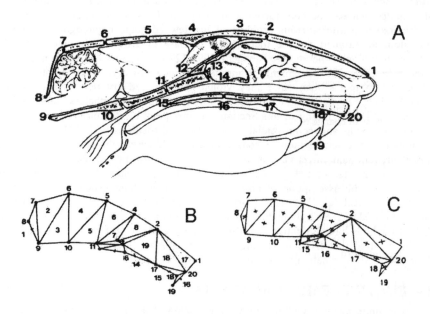

Figure 8.4. The Finite Element method.
(See text for explanation)

As before, FEM does not measure the shape of forms, but only the *shape change between* forms as does BOG. Dealing first with 2-D morphologies, the form to be analyzed with FEM is first subdivided into a series of triangles or 'finite elements' connecting homologous points. Each triangle can be treated as an independent structure except for the sharing of a common side. It is generally assumed that the shape change or growth is constant (or homogeneous) within a triangle. This implies that the change is uniformly distributed across the three homologous landmarks determining the vertices of the triangle. In the words of Richtsmeier and colleagues:

> The form tensor is a numerical description of an ellipsoid, mathematically similar to a variance-covariance or correlation matrix in multivariate statistical analysis. The form tensor identifies orthogonal directions (eigenvectors) of maximum, intermediate and minimum difference between forms at each landmark. These principal directions of strain designate the axes of the ellipsoid. A value (eigenvalue) associated with each direction specifies the length of each axis and thereby defines the local magnitudes of morphological differences along the principal directions (Richtsmeier, *et al.*, 1992:288).

For each triangular element, a 'form difference' tensor is defined as equal to a 'shape difference' tensor plus a 'size difference' tensor. Shape differences are found by subtraction. These shape differences are then averaged over all triangular nodes and over the whole form, to provide summary estimates of change (Cheverud and Richtsmeier, 1986). Or as Richtsmeier, *et al.*, put it:

> Three basic types of information concerning localized morphological differences can be extracted from the form tensor. The size difference, s, measures the increase or decrease in volume required at each landmark to produce the target from the reference form. It is calculated as average of the tensor's eigenvalues, after transformation to a linear scale. Shape difference is composed of two parts, the *magnitude* of shape difference, t, measures the extent of variation in size increase across the anatomical directions (how much the ellipsoid deviates from a sphere) as the standard deviation of the linearly transformed eigenvalues. The *pattern* of shape difference is given by the directions of change of the form tensor (Richtsmeier, *et al.*, 1992:288).

8.4.2. Finite elements in 3-D

In 3-D, the triangles are replaced by hexahedrons (cubes). Each cubic element is composed of eight homologous points in the 3-D (x, y, z) Cartesian system. Again, using an initial form as a base, each cubic element in the base is pairwise compared and the shape change computed as a deformation (Lew and Lewis, 1977: Lewis, *et al.*, 1980; Cheverud, *et al.*, 1983). These cube elements, as the triangular elements in the 2-D case, are independent of each other, with the exception of those surfaces that are joined. These elements are now (in contrast to the 2-D triangles) 'nonhomogeneous'; that is, they represent spatially varying tensor fields. This is also in distinction to BOG where the principal dilations are based on a constant tensor field (Cheverud and Richtsmeier, 1986). Nonhomogeneity offers the advantage of allowing the computation of the local deformation around a landmark. This is not possible in 2-D with the use of triangles, where all three vertices are presumed to have the same deformation since they are, by definition, homogenous. This presumed advantage of FEM over BOG, in particular the property of nonhomogeneity, has been disputed by Bookstein (Bookstein, 1991:251).

Analogous to BOG, if the finite element (hexahedron) departs substantially from a cube becoming thin and elongated, the FEM results will also become increasingly unreliable (Cheverud and Richtsmeier, 1986).

Moss and colleagues have used FEM in 2-space to measure shape changes in the sagittal plane of the growing rat. They fitted triangles to a number of homologous points on the periphery of the rat outline using a crossectional sample composed of five age groups (Moss, *et al.,* 1987). Using FEM, they documented the pattern of cranial growth from 13 to 49 days and demonstrated that the fronto-ethmoidal articulation (element 10) is the primary site of rotation of the facial skull with respect to the cranial base (Fig. 8.4).

The 2-D FEM approach was also utilized to characterize shape changes occurring during the embryological development of the rat cranium, while the 3-D FEM was also used to describe form change in the nasal septum of the same embryonic rat series (Lozanoff and Diewert, 1989; 1993). Other 3-D applications of FEM deal with the numerical description of craniofacial growth in Apert's and Crouzon's syndromes (Richtsmeier, 1987; 1988), the growth of the cranial base in craniosynostosis (Richtsmeier, *et al.,* 1991) and craniofacial growth in primatology (Richtsmeier, 1989).

8.5. THIN PLATE SPLINES

A comparatively new technique is thin plate splines (TPS). Using the theory of surface spline interpolations, (see Bookstein, 1991 for original sources), the development of TPS represents a continuation of Bookstein's work on deformations started with BOG. Bookstein's initial work on BOG (Section 8.3) represents the initial, if limited, approach to put D'Arcy Thompson's transformation grids on a more solid mathematical footing. However, the method of BOG did not provide for a sophisticated graphical display of the point-to-point deformations sought by Bookstein (Reyment, 1991). The following discussion of TPS is intended to briefly convey this method. The method of TPS and the associated approach of 'principal warps' can be found in Bookstein's treatise on landmark data (Bookstein, 1991), which should be consulted for a detailed exposition of the method.

8.5.1. The interpolation Function

TPS is a technique for visualizing form change as a deformation. Involved here, is an interpolation function, which represents a mapping intended to model the 'biological homology' of pairs of points. The interpolant can be thought of as a smooth (well-behaved) function that is fitted to a data point set. In 2-D these could be spline functions such as the Bezier curves. The interpolation function, as a TPS, is then fitted to a surface in 3-D.

8.5.2. Visualization of the Thin Plate Spline

This function can be visualized as a thin metal plate with zero thickness placed over a set of landmarks subject to certain constrains. This surface allows one to visualize the pairwise displacement of landmarks as a deformation. This is done by computing the 'bending energy'. If this plate is completely 'flat' then it has zero bending energy.

Consider a square embedded into some kind of 'elastic' substrate, which is being deformed into a kite by displacing downward the inferior point. Here, three of the four corners are held constant and only the fourth one is displaced directly downward; that is, in such a manner that the x-value remains constant while the y-value decreases (for this example see Bookstein, 1991:31). This downward displacement can be likened to an elastic deformation, which distorts the whole space in which the points are embedded. This deformation in the y-value is computed from a function $z(x, y)$ where the z-coordinate measures 'heights' above and below the surface (thin plate). The larger the deformation, the greater the bending energy and the more 'buckled' the thin plate becomes. Moreover, the plate spline is invariant to rotations, translations and affine transformations, recalling that in the latter case parallel lines remain parallel. To make this clearer, consider the TPS as solely influenced by gravitational forces. That is, even if the thin plate is tilted in 3-D (as long it remains flat), bending energy is not implicated. In contrast, if the thin plate is subject to non-uniform transformations, such as those caused by elastic forces (*e.g.,* stretching or compression) then bending energy is involved which acts to 'wrinkle' the plate in some fashion.

These deformations, viewed as non-uniform transformations, can also be expressed as a sum of the 'principal warps' which are the eigenvectors of the bending energy which correspond to the orthogonal displacements of the landmarks (above and below) from the thin plate (For further discussion of this method the interested reader is referred to Reyment, 1991; Bookstein, 1991).

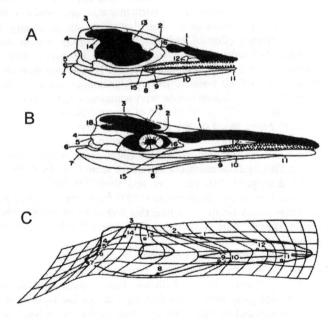

Figure 8.5. The thin plate spline.

Figure 8.5 (adapted from Reyment, 1991) illustrates the TPS applied to: [A] an early marine reptile, the Triassic *ichthyosaur, Grippia,* mapped into [B] the middle Triassic *Cymbospondylus,* form resulting in [C] the total non-affine transformation associated with the TPS. While the thin plate spline can be considered as a sophisticated representation of D'Arcy Thompson's transformation grids, it provides little insight regarding the biological processes that are responsible for the observed changes.

The deformation methods reviewed here, BOG, FEM and TPS, can be considered as novel contributions in that they: [1] represent sophisticated mathematical methods for numerically describing the biological form; [2] attempt to deal with the deficiencies inherent in CMA; and [3] represent the foundation of what will undoubtedly be an entirely new direction to morphometrics in the future. All these methods, however, also share one aspect with CMA, which is the dependency on homologous points. A dependency, as indicated earlier, that is due, in part, to historical tradition.

8.6. EUCLIDEAN DISTANCE MATRIX ANALYSIS

Finally, another comparatively recent coordinate-free method remains to be mentioned, which is Euclidean Distance Matrix Analysis (EDMA). EDMA was developed by Lele, a statistician at John Hopkins (Lele, 1993; Lele and Richtsmeier, 1990; 1991; 1992; Richtsmeier, *et al.,* 1991). Recall that one of the properties of a desirable form representation is invariance to coordinate transformations such as uniform scaling, translation, rotation and reflection (Section 6.3.1). Such a representation is possible with EDMA (Lele, 1991; Richtsmeier, *et al.,* 1992). In brief, EDMA utilizes 3-D Cartesian coordinates of the homologous points to identify local areas of significant shape change (Corner and Richtsmeier, 1993). As Euclidean distances among all possible landmarks are involved, the method becomes independent on the Cartesian co-ordinate system. Three steps are involved: [1] the computation of a 'mean form' from the distances of all possible pairs of landmarks (these individual distances, viewed as an individual Euclidean matrix representation, are then averaged to yield a mean matrix for the sample); [2] the calculation of a form *difference* matrix using a ratio of similarity between forms based on the individual pairwise distances between two mean forms being compared; and [3] this distance difference matrix is then sorted to identify the areas of maximum and minimum change. Local areas of maximum and minimum change are identified by how much the difference matrix values differ from one. These ratios do not distinguish between 'size and shape' as both aspects are included in the analysis. See Corner and Richtsmeier (1993) for a recent application of EDMA to the primate craniofacial complex. While this technique is independent of FEM, results are generally similar. This is perhaps not unexpected if both approaches are based on the same homologous-point data set.

Because of the landmark-based constraint, none of these approaches can be considered as complete models of form in themselves. While they can be considered as a successful mapping of homologous points between forms, they exclude other information present in the biological form. This missing information consists of the boundary outline between points and the structural or textural information lying within the bounded outline. The next chapter takes up the first element of this missing information; namely, the bounded outline of the biological form.

KEY POINTS OF THE CHAPTER

This chapter has outlined a number of analytic procedures that fall under the category of coordinate morphometrics. These methods focused on points, or more specifically, homologous landmarks, in either 2-D or 3-D. Three of the methods, BOG, FEM and TPS are considered deformation methods that measure changes in form rather than the form itself. All of the methods can be considered as alternatives that attempt to overcome the limitations inherent in CMA, that is, conventional metrics consisting of distances, angles and ratios.

CHECK YOUR UNDERSTANDING

1. Trace the development of the conventional metrical approach (CMA) from the time of Camper.

2. Give some examples were conventional metrics (CMA) can be considered perfectly appropriate and some were they are not?

3. Why have angles been considered as measures of shape? Why are such measures insufficient?

4. Besides the deficiencies listed in the chapter, can you think of any other limitations that are inherent in the use of conventional metrics?

5. Why have conventional metrics, consisting of distances, angles and ratios, become so predominant in the biological sciences, in spite of their deficiencies?

6. What is the difference between point homology and structural homology? Can you think of ways that these two views can be reconciled?

7. Why have ratios been considered as measures of shape? What objections can you raise to the use of ratios as measures of shape?

8. Why do you have to be cautious with the use of ratios in biology? Find some biological examples of actual ratios to bolster your position.

9. What are the differences between BOG and FEM? Do these differences have an effect on the interpretation of biological shape changes?

10. What exactly is meant by the statement: 'BOG measures shape change not shape itself'?

11. Why are the techniques of BOG, FEM and TPS called form deformation methods?

12. Utilizing the principal dilations in BOG, how are size and shape changes interpreted? How are such form changes handled with FEM?

13. How are size and shape changes carried out in EDMA?

REFERENCES CITED

Albrecht, G. H. (1978) Some comments on the use of ratios. *Syst. Zool.*, **27**:67-71.

Albrecht, G. H., Gelvin, B. R. and Hartman, S. E. (1993) Ratios as size adjustment in morphometrics. *Am. J. Phys. Anthrop.* **91**:441-468.

Atchley, W. R. (1978) Ratios, regression intercepts, and the scaling of data. *Syst. Zool.* **27**:78-83.

Atchley, W. R., Gaskins, C. T. and Anderson, D. (1976) Statistical properties of ratios. I. Empirical results. *Syst. Zool.* **25**:137-148.

Bathe, K. J. and Wilson, E. L. (1976) *Numerical Methods in Finite Element Analysis.* Englewood Cliffs, New Jersey: Prentice-Hall.

Bookstein, F, L. (1977) Orthogenesis of the hominids: An exploration using biorthogonal grids. *Science* **197**:901-904.

Bookstein, F, L. (1978) The measurement of biological shape and shape change. Lecture notes in biomathematics. New York: Springer-Verlag **24**:1-191.

Bookstein, F, L. (1982) On the cephalometrics of skeletal change. *Am. J. Orthod.* **82**:177-182.

Bookstein, F, L. (1983) Geometry of craniofacial growth invariants. *Am. J. Orthod.* **83**:221-234.

Bookstein, F, L. (1984) A statistical method for biological shape comparisons. *J. Theor. Biol.* **107**:475-520.

Bookstein, F, L. (1986) Size and shape spaces for landmark data in two dimensions. *Stat. Sci.* **1**:181-242.

Bookstein, F, L. (1991) *Morphometric Tools for Landmark Data.* Cambridge: Cambridge University Press.

Cartmill, M. (1994) A critique of homology as a morphological concept. *Am. J. Phys. Anthrop.* **94**:115-124.

Cheverud, J. M. and Richtsmeier, J. T. (1986) Finite-element scaling applied to sexual dimorphism in rhesus macaque (*Macaca mulatta*) facial growth. *Syst. Zool.* **35**:381-399.

Cheverud, J. M., Lewis, J. L., Bachrach, W. and Lew, W. D. (1983) The measurement of form and variation in form: An application of three-dimensional quantitative morphology by finite-element methods. *Am. J. Phys. Anthrop.* **62**:151-163.

Corner, B. D. and Richtsmeier, J. T. (1993) Cranial growth and growth dimorphism. *Am. J. Phys. Anthrop.* **92**:371-394.

Corruccini, R. S. (1973) Size and shape in similarity coefficients based on metric characters. *Am. J. Phys. Anthrop.* **38**:743-754.

Corruccini, R. S. (1975) Multivariate analysis in biological anthropology: Some considerations. *J. Hum. Evol.* **4**:1-19.

Corruccini, R. S. (1977) Correlation properties of morphometric ratios. *Sys. Zool.* **26**:211-214.

Dryden, I. L. and Mardia, K. V. (1998) *Statistical Shape Analysis*. New York: John Wiley and Sons.

Dodson, P. (1978) On the use of ratios on growth studies. *Syst. Zool.* **27**:62-67.

Garn, S. M., Holly Smith, B. and La Velle, M. (1984) Applications of pattern profile analysis to malformations of the head and face. *Radiol.* **150**:683-690.

Hills, M. (1978) On Ratios—A response to Atchley, Gaskins, and Anderson. *Syst. Zool.* **27**:61-62.

Huiskes, R. and Chao, E. Y. S. (1983) A survey of finite element analysis in orthopedic biomechanics: The first decade. *J. Biomech.* **16**:385-09.

Kaesler, R. L. (1967) Numerical taxonomy in invertebrate paleontology. **In** *Essays in Paleontology and Stratigraphy*. Teichert, C. and Yochelson, E. L. (Eds.) Lawrence, Kansas: University of Kansas Press.

Lele, S. (1991) Some comments on coordinate free and scale invariant methods in morphometrics. *Am. J. Phys. Anthrop.* **85**:407-417.

Lele, S. (1993) Euclidean distance matrix analysis (EDMA): Estimation of mean form and mean form difference. *Math. Geol.* **25**:573-602.

Lele, S. and Richtsmeier, J. T. (1990) Statistical models in morphometrics: Are they realistic? *Syst. Zool.* **39**:60-69.

Lele, S. and Richtsmeier, J. T. (1991) Euclidean distance matrix analysis: A coordinate free approach for comparing biological shapes using landmark data. *Am. J. Phys. Anthrop.* **86**:415-427.

Lele, S. and Richtsmeier, J. T. (1992) On comparing biological shapes: Detection of influential landmarks. *Am. J. Phys. Anthrop.* **87**:49-66.

Lestrel, P. E. (1974) Some problems in the assessment of morphological size and shape differences. *Yearbook Phys. Anthrop.* **18**:140-62.

Lestrel, P. E. (1989) Method for analyzing complex two-dimensional forms: Elliptical Fourier functions. *Am. J. Hum. Biol.* **1**:149-64.

Lew, W. D. and Lewis, J. L. (1977) An anthropometric scaling method with application to the knee joint. *J. Biomech.* **10**:171-181.

Lewis, J. L., Lew, W. D. and Zimmerman, J. R. (1980) A nonhomogeneous anthropometric scaling method based on finite element principles. *J. Biomech.* **13**:815-824.

Lozanoff, S. and Diewert, V. M. (1989) A computer graphics program for measuring two- and three-dimensional form change in developing craniofacial cartilages using finite element methods. *Comp. Biomed. Res.* **22**:63-82.

Lozanoff, S., Zingeser, M. R. and Diewert, V. M. (1993) Computerized modelling of nasal capsular morphogenesis in prenatal primates. *Clin. Anat.* **6**:37-47.

Mardia, K. V. and Dryden, I. L. (1989) The statistical analysis of shape data. *Biometrika* **76**:271-281

Medawar, P. B. (1945) Size, shape and age. In *Essays on Growth and Form Presented to D'Arcy Wentworth Thompson*. Le Gros Clark, W. E. and Medawar, P. B. (Eds.) Oxford: Clarendon Press.

Moss, M. L. and Young, R. W. (1960) A functional approach to craniology. *Am. J. Phys. Anthrop.* **18**:281-292.

Moss, M. L., Vilmann, H., Moss-Salentjin, L., Sen, K., Pucciarelli, H. M. and Skalak, R. (1987) Studies on orthocephalization: Growth behavior of the rat skull in the period 13-49 days as described by the finite element method. *Am. J. Phys. Anthrop.* **72**:323-342.

O'Higgins, P. (1989) Developments in cranial morphometrics. *Folia Primatol.* **53**:101-124.

O'Higgins, P. and Johnson, D. R. (1988) The quantitative description and comparison of biological forms. *Critical Rev. Anat. Sci.* **1**:149-170.

Read, D.W. (1990) From multivariate to qualitative measurement: Representation of shape. *Hum. Evol.* **5**:417-429.

Reyment, R. A. (1991) *Multidimensional Paleobiology*. Oxford: Pergamon Press.

Richtsmeier, J. T. (1987) Comparative study of normal, Crouzon and Apert craniofacial morphology using finite element scaling analysis. *Am. J. Phys. Anthrop.* **74**:473-493.

Richtsmeier, J. T. (1988) Cranial growth in Apert syndrome as measured by finite element scaling analysis. *Acta Anat.* **133**:50-56.

Richtsmeier, J. T. (1989) Applications of finite-element scaling analysis in primatology. *Folia Primatol.* **53**:50-64.

Richtsmeier, J. T., Grauz, H. M., Morris, G. R., Marsh, J. L. and Vannier, M. W. (1991) Growth of the cranial base in craniosynostosis. *Cleft. Pal. Cranio. J.* **28**:55-67.

Richtsmeier, J. T., Cheverud, J. M. and Lele, S. (1992) Advances in anthropological morphometrics. *Ann. Rev. Anthropol.* **21**:283-305.

Simpson, G. G. (1961) *Principles of Animal Taxonomy*. New York: Columbia University Press.

Thompson, D. W. (1915) Morphology and mathematics. *Trans. Roy. Soc. Edinburgh* **50**:857-895.

Thompson, D. W. (1942) *On Growth and Form*. (2nd Ed.) Cambridge: Cambridge University Press.

Tobias, P. V. (1959) Studies on the occipital bone in Africa: I. Pearson's occipital index and the chord-arc index in modern African crania: means, minimum values and variability. *J. Roy. Anthrop. Inst.* **89**:233-252.

White, R. J. Phenice, H. C. (1988) Comparison of shape description methods for biological outlines. **In** *Classification and Related Methods in Data Analysis*. Bock, H. H. (Ed.) North Holland: Elsevier Science Pub. BV.

9. BOUNDARY MORPHOMETRICS

Nessuna humana investigazione si pio dimandare vera scienzia s'essa non passa per le matimatiche dimostrazione.
No human investigation can be called real science if it cannot be demonstrated mathematically.

Leonardo da Vinci (1452-1519)

9.1. INTRODUCTION

The methods to be described here focus on the boundary of outlines, rather than on landmarks. These approaches become increasingly useful when there are only a few, sparsely located, homologous landmarks, and are essential in the absence of such points. These methods become indispensable if the global aspects of the boundary outline, as a totality in itself, are of interest. This latter point is predicated on the assumption that a majority of the biological information of interest (especially with respect to shape) is confined to the outline. Visual information that is found within the outline is considered structural (or textural) in nature and will be taken up in Chapter 10.

Fourier descriptors (FDs) are particularly useful in analyzing outline information as well as are elliptical Fourier functions (EFFs).[75] These approaches will be described in some detail in Sections 9.3, 9.4 and 9.5. Ultimately, what will be needed is the ability to relate the functional or behavioral correlates that make up the actual form, with mathematical parameters describing the form. Ideally, this implies the extraction of biological meaning from the coefficients that define the function under consideration. However, this ability to relate such biological parameters to the observed form has proven to be quite difficult in practice.

One of the advantages of these boundary morphometric methods is that one can generally precisely re-create the boundary from the function. This is an *information-preserving* property. Attempts have been made to extend some of the outline methods to 3-D. Because of the complexity of most biological forms, an overwhelming number of studies continue to be limited to 2-D. It is to be emphasized that point homology, as defined with the use of landmarks (discussed in Section 8.2.1), is generally lost with these boundary methods. However, a recent development has allowed: [1] a limited extension of EFFs to 3-D and [2] the incorporation of homologous data points. This approach will be presented in Section 9.5. While both open curves as well closed curves can be analyzed, the majority of studies have been on closed curve data sets so this review is largely limited to closed curves, although a recent study of the dental arch form provides an example of open curves (Ahn, *et al.*, 1999; Lestrel, *et al.*, 1999; Takahashi, *et al.*, 1999). A discussion of such open curves can also be found in Rohlf (1990).

---

[75] See Lestrel (1997b) for a more extensive discussion of Fourier descriptors in boundary morphometrics, as well as numerous applications in a biological context.

Six boundary morphometric methods, which have received considerable attention in the literature, will be examined. These are median axis functions, FDs, eigenshape analysis, EFFs, Fourier transforms and wavelets. These methods generally start with a set of closely ordered (*x, y*) coordinates, which are either evenly spaced (characteristic of most studies) or unevenly distributed along the boundary outline of the form. The latter case allows flexibility, as regions of sharp curvature may need more points to accurately describe the outline. Since these points are closely spaced, they are also highly correlated with each other, so it is desirable to reduce this comparatively large number to a smaller, more manageable number of variables, which are, ideally, uncorrelated. To this end the use of FDs are particularly suitable because they are composed of trigonometric functions, which generate orthogonal coefficients (Tolstov, 1962). Once this smaller set of variables has been computed (*e.g.*, harmonic coefficients or amplitudes) they can be subjected to further analysis using multivariate methods such as discriminant functions or principal components as described earlier in Chapter 7.

9.2. MEDIAN AXIS TECHNIQUES

The first method to be considered that deals with the boundary of the form, in contrast to landmarks, is the median axis function (MAF). This method is known variously as the median axis, medial axis (Lee, 1982), symmetric axis (Blum, 1967), or line skeleton (Bookstein, 1979). MAF approaches can be considered as coordinate-free methods in the sense that they are generally not dependent on homologous points, although homology could be presumably imposed (but see Section 9.2.2). The use of the MAF can be viewed as global in the sense that the method makes use of the total outline instead of focusing on localized landmarks, which invite the subjectivity that is characteristic of CMA.

9.2.1. Definition of the Median Axis

The MAF is defined as the locus of points, which lie in the interior of the form exactly equidistant from the border of the outline. MAF was developed by Blum as a way of simplifying planar morphologies (Blum 1967; 1973; Montanari, 1969; Blum and Nagel, 1978). This procedure allows the collapsing of a 2-D outline onto a curve. The approach consists of embedding a series of overlapping circles (also called disks) that touch the outline in such a way that they are tangential to the borders of the form. The centers of each of these circles define a point on the MAF. These points, or nodes, can then be connected with a curve to create, what is sometimes termed, a 'stick figure'. An alternative way in which the MAF has been viewed is think of the form collapsing into itself in a direction normal to the boundary (Oxnard, 1973). Figure 9.1 is an example of the median axis approach applied to a human mandible. The stick figure has two bifurcations here, one within the ramus and the other anteriorly near the mental symphysis.

The MAF then, is produced from the loci of these points. Branches or bifurcations can also appear as in Figure 9.1. The procedure requires that a circle be totally contained within the region of interest, but not totally contained within any other circle. Note here that a circle of maximum radius will touch the outline in at least two places. If the circle touches at more than two places on the outline, then that circle also encloses a bifurcation. The radius of each circle determines the 'width' of the MAF at those tangent points. Thus, there are two parts to the technique: [1] the curve that defines the MAF; and [2] the radius function, R, which is defined by the circle radii at each of the point tangents on the boundary. MAF can be thought of a primitive descriptor of the shape. Primitive in the sense that it represents a reduction to a smaller set of variables composed of two parts, which together contain all the information needed to re-create the form. To make the visualization of the MAF more apparent, consider the following simple planar figures: [1] for a circle the MAF is a point, [2] for an ellipse the MAF will be displayed as a line, and [3] for a rectangle the MAF will contain two bifurcations resulting in five lines (Daegling, 1993).

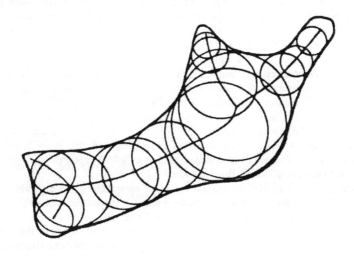

Figure 9.1. Median axis applied to a human mandible.
(See text for explanation)

9.2.2. Variations on a Theme

There seem to be at least five distinct MAF algorithms in use, which produce different results. See Pavlidis (1978) for an early review of this method and other techniques applied in a pattern recognition context. For purposes of discussion here, 'median axis' or MAF, will be used as the general term for all these procedures. Although Blum and others viewed the MAF as a curve, it is possible to also approximate it with line segments connecting the centers of the inscribed circles. Such an algorithm that computes the *discrete* MAF (in contrast to the continuous case developed by Blum) was derived by Bookstein (1979). This approximation depends on the geometric construction of certain pseudo-normals; that is, lines computed normal to the boundary outline at points located on the boundary. This pseudo-normal can be also thought of as an approximation for the radius function, R, discussed earlier. This MAF, defined as a line skeleton, would be appropriate for those cases where the boundary outline is sparsely composed of points. This is the situation that would inevitably arise if one strictly adhered to the requirement of homologous landmarks (Bookstein, 1979; Daegling, 1993). Curiously, the illustrated mandibular example that Bookstein uses shows the MAF with two bifurcations yielding six lines, with an extra line radiating toward the *gonial* aspect[76] (not five as in Fig. 9.1). This is in contrast with the mandibular example (see below), discussed by Webber and Blum (1979) where the two bifurcations lead to five lines (as in Fig. 9.1). The reason for this anomaly between the two MAF algorithms is not readily apparent.

Lee (1982) provided an improved algorithm in terms of computer speed that computes the MAF of a planar shape represented by an n-sided simple polygon. Most of the MAF computations require time proportional to n^2 where n is the number of boundary edges of the figure. Lee's algorithm, in a comparable way to the FFT (Cooley and Tukey, 1965), takes $n \log n$ time (the FFT will be encountered again in Section 9.6). Montanari (1969) is an early example of a MAF algorithm applied to chromosome data.

The MAF approach has been generalized to 3-D by using spheres instead of circular disks (O'Rourke and Badler, 1979; Borgefors, *et al.*, 1999). In an analogous sense to the 2-D case, a simple figure like a sphere would have a point as the MAF. A parallelepiped (a box) would have a MAF with nine lines. It is cautioned that although the generalization to 3-D in this case seems straightforward, this is not always the case. For example, the properties of area and perimeter of a 3-D object do not seem to remain invariant with the use of MAF (Blum, 1973).

Webber and Blum (1979) applied the MAF approach to the mandible in 2-D and discovered that a stable center seemed to be present at the intersection of the medial axis defining the body of the mandible and the bifurcation that led to the tip of the coronoid process and to the condyle head. Further, the three angles defining this 'center' also showed considerable stability in a normal mandibular series. The presence of these bifurcations that form intersections may lend themselves as 'internal' landmarks, which could be subsequently used for mandibular superimposition. They present such an example with mandibles displaying the Treacher-Collins syndrome. [77] When these mandibles are compared with a normal series, marked departures from the normal angular values could be discerned. Nevertheless, the implication that these 'internal' landmarks, no matter how useful, are homologous can be questioned. Strictly speaking, these are constructed points, which are derived from the boundary outline and do not readily display the required 1:1 mapping into another form.

Straney (1990) utilized the MAF approach to the study of evolutionary changes in shape of certain bones (the *baculum*) in the spiny rat (genus *proechimys)*. Previous studies had been based on length and width measurements because of the lack of

---

[76] The gonial aspect refers to where the inferior body of the mandible meets the posterior ascending ramus.
[77] Treacher-Collins is thought to be inherited as an autosomal dominant with variable penetrance (chromosome 5). Approximately 50-60% of cases are presumed to be new mutations. Syndrome is characterized by facial bone hypoplasia, breathing, speech and hearing difficulties. Surgical treatment generally is required.

homologous landmarks. Points on the outline (from 150-350) were digitized and Bookstein's line skeleton applied. The question that arose was how effective was the line skeleton and its branch points (bifurcations) for capturing the shape variation in the outline. Using the bifurcation intersections as points, a set of twenty-seven inter-landmark distances were computed. To these were added ten variables computed from the radius or R function of the line skeleton. These thirty-seven variables were then submitted to a principal component analysis (PCA). The plot of PC-1 against PC-2 displayed reasonably good discrimination among most of the six groups of *Proechimys*. At this juncture, a second question arose; namely, since the line skeleton was a simplification of the actual shape, how much does the simplification affect group discrimination? To test this, the bacular outlines were also analyzed by Rohlf (1990) using EFFs (Section 9.5). The accuracy of the EFF fit with 22 harmonics, in terms of residuals, was not given. Size-standardization with the EFF was not imposed in this case to maintain conformity with the line skeleton data. The PCA results of the EFF coefficients (the actual coefficients, not their amplitudes) were essentially similar to the earlier line skeleton data after a number of transformations so that both approaches were comparable. For example, the PCA axes were standardized to have unit variance. Size, in this case, played a major part in the group discrimination with the use of EFFs and this is, in turn, affected the PC-1 axis.

An attempt to describe the variation in the ape mandibular symphysis using MAF, demonstrated that while the line skeleton was useful as a simplifying technique for the study of shape, it was not very useful for taxonomic purposes because of the extensive variability present (Daegling, 1993). Daegling suggested that a direct analysis of the outline using EFFs might be more useful for describing the symphysis shape in these particular primate groups.

While MAF methods may be potentially useful as descriptors of shape, a number of problems have arisen with their use. At least for simple forms, the MAF is not unique. For example, it is possible for classes of similar shapes to have identical median axis representations. Consider the simple example of two ellipses, both of them with equal major axes but one of them with a minor axis twice as large as the other. In this case, the MAF will be identical in the two forms. It is for this reason that the radius function must also be computed. Another issue, which has hindered general acceptance of the technique, is its sensitivity. That is, the sensitivity of the method to noise can be unduly large (Blum and Nagel, 1978; Pizer, *et al.*, 1987). Subtle changes in the form can also result in a discontinuous change in the MAF representation. What this means is that minor perturbations in the outline, *i.e.*, small inflections, can result in bifurcations in one form and not in another as mentioned earlier. Consequently, such small perturbations can lead to classification errors. An attempt to correct this situation was an approach described as 'hierarchical' (Pizer, *et al.*, 1987). That is, by initially viewing the form in global terms and then breaking the form down into its components in terms of scale. The idea behind this approach was that the smaller the component, *e.g.*, a specific bifurcation, the less its impact. The approach consists of focusing on the bifurcations of the MAF and systematically removing them, in effect a smoothing technique. However, over-correction could result in the elimination of components that might have diagnostic value in the application of this technique to biological data. This raises the theoretical question of how to determine the smoothing threshold. Finally, it is difficult to identify the biological processes responsible for the presence of the bifurcations or centers. Further work is needed.

9.3. CONVENTIONAL FOURIER DESCRIPTORS

Fourier analytic techniques continue to be extensively used in very diverse fields (Lestrel, 1997b). They also continue to spawn new developments, *e.g.,* wavelets. Thus, one can trace a definitive historical development over the last three decades, that started with conventional Fourier descriptors (FDs), led to the continuous Fourier transform (FT), followed by the discrete Fourier transform (DFT), then the fast Fourier Transform (FFT), the short time Fourier transform (STFT), and finally to the current work on wavelets; specifically, the discrete wavelet transform (DWT). These developments will be further examined in Sections 9.6 and 9.7.

The approach of Fourier analysis can be viewed as a transformation of data from one domain to another. In electronic engineering, this transformation is from the *time* domain into the frequency domain. In other fields that commonly use FDs, such as pattern recognition and the geological and biological sciences, the transformation is generally from the *spatial* domain, rather than time, into the frequency domain. The spatial domain refers to the data points that determine the outline or boundary of the planar object of interest and the frequency domain is characterized in terms of a new set of variables describing amplitude and phase relationships. This has also been termed a decomposition of the spatial configuration (boundary) into frequency components (amplitude and phase). This procedure of decomposition is known as *harmonic analysis* or Fourier analysis. A useful property of FDs is that they can be used to re-create the outline of the form under consideration (*information-preserving* property*)*, in contrast to landmark methods. The converse process of re-creation is called *harmonic synthesis*. Useful discussions can be found in Tolstov (1962), Kline (1972) and Davis (1986).

Two FD approaches have been widely used. Both convert the data to polar coordinates, prior to Fourier analysis. One is based on measurements from a center within the form, preferably the centroid. The other uses an angular function based on the points located on the outline (taken up in Section 9.4).

9.3.1. Frequency, Amplitude and Phase Relationships

Three aspects of Fourier analysis deserve comment, these are: the *period*, the *amplitude* and the *phase*. Consider a simple sinusoidal function such as a sine wave, which repeats over an interval (along the x-axis). The period, L, refers to one complete cycle from 0 to 2π radians or 360 degrees. The frequency, on the other hand, is the reciprocal of this period or wavelength, that is, $f = 1/L$. The fundamental frequency is given by the first harmonic, which is, generally equal to the interval from 0 to 2π or $[-\pi, \pi]$. The next higher frequency is the second harmonic, which is one-half the wavelength of the fundamental frequency, and so on. The amplitude, in the Cartesian co-ordinate system, refers to the maximum height of the waveform from the x-axis (measured along the y-axis). Phase refers to the displacement of the starting point of the waveform from the origin. For example, the form of the sinusoidal curve, $y = cos\ x$, is identical to $y = sin\ x$, except that it is shifted in phase by 90 degrees or $\pi/2$. Thus, these three properties:

period, *amplitude* and *phase* provide for a flexible system that can be used to fit many forms. The spatial form under consideration is made 'periodic' in the sense that it is repeated over a set interval (the period), usually from *0* to *2π*. That is, the last data point on the outline is followed by the first point, and the process is repeated (Lestrel, 1980; 1982).

Finally, it should be mentioned, that both 1-D and 2-D FDs are used. The 1-D FD can be used to analyze closed boundary outlines (both in polar and Cartesian coordinates), while 2-D FDs, set up as a double Fourier series, are limited to the Cartesian system with sinusoidal waves extending in both the *x*- and *y*-directions (along the *xy*-plane), with the amplitude measured along the *z*-axis (Lestrel, 1980).

9.3.2. Fourier's Series

Before examining Fourier's series, it might prove useful to provide a non-rigorous development for the solution of a series. Recalling that the equation of a straight line associated with linear regression, can be represented by:

$$y = a + bx, \qquad [9\text{-}1]$$

where the constant, *a*, refers to the *y*-intercept and the coefficient, *b*, is the slope of the line. The slope can also be computed from the angle that the line makes with the x-axis. Evaluation of the *a* and *b* values is via the procedure of least squares (Draper and Smith, 1998). If we add a third term, cx^2, we arrive at a parabola:

$$y = a + bx + cx^2. \qquad [9\text{-}2]$$

The addition of more terms leads to an *n*th order polynomial of the form:

$$y = a + bx + cx^2 + dx^3 + \cdots + kx^n. \qquad [9\text{-}3]$$

Applying the same procedure using cosine terms results in:

$$y = a_0 + a_1 \cos x, \qquad [9\text{-}4]$$

where a_0 is a constant. Extending this procedure in a similar way to *n*th order polynomials yields:

$$y = a_0 + a_1 \cos x + a_2 \cos 2x + a_3 \cos 3x + \cdots + a_n \cos n x. \qquad [9\text{-}5]$$

Equation [9-5] can be shortened using the summation, Σ, symbol:

$$y = a_0 + \sum_{n=1}^{k} a_n \cos nx, \qquad\qquad [9\text{-}6]$$

where k refers to the total of all cosine terms.[78] In an analogous way, a sine series can also be constructed:

$$y = \sum_{n=1}^{k} b_n \sin nx. \qquad\qquad [9\text{-}7]$$

The summation of equations [9-6] and [9-7] together constitutes Fourier's series:

$$y = a_0 + \sum_{n=1}^{k} a_n \cos nx + \sum_{n=1}^{k} b_n \sin nx \qquad\qquad [9\text{-}8]$$

Note: The nx term here is simplified by making the amplitude $\varpi = 1/T$ and the fundamental frequency, T, equal to the period of 2π (Tolstov, 1962; Lestrel, 1997b).

9.3.3. Fourier's Series as Discrete Approximations

If the observed data can be described with a set of discrete points (x, y) on the boundary outline (generally equally-spaced), then a morphological form can be represented as a tabulated function.[79] Again, if we define the period over a 2π interval, equation [9-8], we can display the familiar discrete[80] form of Fourier's series in Cartesian coordinates as:

$$f(x) = a_0 + \sum_{n=1}^{k} a_n \cos nx + \sum_{n=1}^{k} b_n \sin nx, \qquad\qquad [9\text{-}8]$$

where a_0 is the constant, or mean, a_n and b_n are the Fourier coefficients, n is the harmonic number and x refers to the points sampled over the period along the x-axis, and k is the maximum harmonic number. The a_0, a_n, and b_n coefficients are computed from the following equations.[81]

---

[78] It is to be assumed here, without formal proof, that these series must be convergent. A convergent series has to lead to a finite sum (a solution) which is in contrast to a divergent series which does not lead to a finite sum and hence, no solution.

[79] As one usually is dealing with a set of observed data points on the periphery of the boundary, a function such as $y = f(x)$ representing the data is initially unknown. Thus, one has to start with a 'tabulated function', which is set up as a table, in this case consisting of the (x, y) coordinates of the boundary points.

[80] The discrete formulation has definite limits (as set by the summation), in contrast to the continuous case, where the limits of the integrals may range from $-\infty$ to $+\infty$.

[81] Fourier's series is set up here as a discrete periodic function, which requires the use of numerical integration methods to solve for the coefficients; hence, the summation signs in equations [9-9] through [9-11]. In contrast, continuous periodic functions require integration, in which case equations [9-9] through [9-11] would be written with integral signs.

Letting the sampling along an outline contain k observations with even divisions over the interval $[-\pi, \pi]$ then:

$$a_0 = \sum_{n=0}^{k-1} f(x)/k, \qquad\qquad\qquad [9\text{-}9]$$

$$a_n = \frac{2}{k} \sum_{n=0}^{k-1} f(x)\cos nx, \qquad (n = 0,1,2,\cdots,\frac{k-1}{2}) \qquad [9\text{-}10]$$

and

$$b_n = \frac{2}{k} \sum_{n=0}^{k-1} f(x)\sin nx, \qquad (n = 1,2,\cdots,\frac{k-1}{2}) \qquad [9\text{-}11]$$

where n is the harmonic number, k is the maximum harmonic, subject to Nyquist frequency requirements. The Nyquist or folding frequency is the maximum frequency that can be detected from the sampled data. This is necessary to avoid distortion or 'aliasing' of the signal. Thus, the maximum number of harmonics cannot exceed ½ the number of sampled data points (Newland, 1993). The limits $n=0$ and $n=k-1$ follow from the trapezoidal rule (Harbaugh and Merriam, 1968; Lestrel, 1997b).

If the tabulated function can be expressed as set of distances to a set of points on the outline from a pre-determined center within the closed form, the Fourier series can then be defined in polar coordinates (r_i, θ_i) as:

$$f(\theta) = A_0 + \sum_{n=1}^{k} a_n \cos n\theta + \sum_{n=1}^{k} b_n \sin n\theta, \qquad [9\text{-}12]$$

where the period is again defined over a 2π interval and θ is in radians. The maximum degree or harmonic number is denoted by k, while a_n and b_n are the Fourier coefficients for the nth harmonic (Lestrel, 1974; 1980; 1982; 1997a; Parnell and Lestrel, 1977). Regardless of whether the function is embedded in the polar or Cartesian system, one can compute a new variable utilizing both coefficients (a_n and b_n). This is called the amplitude and computed for each harmonic number. The amplitude is defined as:

$$A_n = \sqrt{a_n^2 + b_n^2}, \qquad\qquad\qquad [9\text{-}13]$$

where A_n is the amplitude for the nth harmonic, a_n and b_n are the Fourier coefficients, and n is the harmonic number. From the amplitude, one can derive the *power*, which is defined as:

$$P_n = \frac{A_n^2}{2} = \frac{a_n^2 + b_n^2}{2}.$$ [9-14]

Utilizing equation [9-13], one can plot the *amplitude spectrum* as the amplitude versus harmonic number n. Similarly, from equation [9-14], one can plot the *power spectrum* as a graph of power versus harmonic number. It is of interest that the power spectrum is also a description of the contribution that each harmonic makes in terms of the explained variance (Davis, 1986; Lestrel, 1997a).

The remaining property, phase, deserves special attention. The phase angle for the nth harmonic is computed as:

$$\Phi_n = \tan^{-1}\left[\frac{b_n}{a_n}\right],$$ [9-15]

where the a_n and b_n are again the coefficients for the nth harmonic. Many of the FD studies have ignored phase and emphasized only differences in the amplitudes. Others have suggested that the phase angle information cannot be ignored (Rohlf, 1986; Kaesler, 1997). This is technically correct if the outlines differ in orientation; for example, with differing positions for the starting angle and its radius from the center, comparisons between outlines may be biased. However, problems with significantly differing initial phase angles can be minimized with a carefully chosen common orientation of the outlines *prior* to Fourier analysis. Ehrlich and co-workers have also demonstrated how the phase angle 'locks' onto homologous features suggesting invariance of the phase angle within samples of similar specimens (Ehrlich, *et al.*, 1983). This can be considered as *prima facie* evidence that phase and amplitude are largely independent of each other.

9.3.4. Residuals, Positional-orientation and Size-standardization

Before we take a closer look at FDs and their application to data, three specific aspects need to be considered. One of the practical questions that arise with the use of FDs (and also with EFFs discussed in Section 9.5) has to do with *goodness-of-fit* requirements. That is, how many harmonics are really needed to achieve a satisfactory fit of the FD as an expected function to the boundary outline (observed form). The answer is found by computing the *residual* or difference between the points derived from the FD and the points defining the observed data. Using 15 to 30 harmonics, yielded mean residual values (averaged over the whole form) in the range of 0.05 to 0.20 mm, which was found to be satisfactory for a wide variety of data sets.[82]

Besides the residual, there are two normalizations termed *positional orientation* and *size-standardization* to be considered (Parnell and Lestrel, 1977; Lestrel, 1980). Positional-orientation refers to two aspects: [1] the orientation of the outline in space

---

[82] Besides the residual, it should be noted that, in the collection of outline data, scaling plays a part. Large morphological images are easier to handle than very small ones, in terms of the errors of measurement.

(the issue of phase) and [2] the placement of the center from which the vectors to the outline are constructed and measured (assumed to be in polar coordinates). The first aspect can be controlled with judicious positioning of the outline prior to digitization. The second aspect requires that a neutral center, the centroid, be computed. The centroid is the only center from which the position of the vectors remains invariant to rotation. To achieve this, a recursive process is required to insure that the vector center is translated to the centroid. This requires a re-computation of the vector lengths to the outline. The net effect of this procedure is to drive the a_1 coefficient to zero as well as to alter the values the higher harmonic coefficients $(a_2 \cdots a_n)$. This recursive procedure was developed in Parnell and Lestrel (1977). Thus, the use of any other center will result in increased variability in the Fourier coefficients (in a sense 'noise') due strictly to the position of the center, and *not* to the variability in the outline (Full and Ehrlich, 1982). The development of EFFs makes conventional FDs somewhat outdated (Section 9.5).

Finally, at times, it may be judicious to impose symmetry on an outline that is an open curve; that is, *not* bounded (see examples in Sections 7.2.4 and 9.3.5). Symmetry, or reflection, can often be imposed by creating a mirror-image about an axis. This has a two-fold advantage in that: [1] the outline can be bounded and [2] it leads to a 50% reduction of the number of computed Fourier coefficients. The reason for the reduction in coefficients is that sine terms measure asymmetry and cosine terms measure symmetry. Since the bounded outline has now been made symmetric, the sine terms will vanish leaving a function solely composed of cosine terms (Lestrel, 1980; 1997).

The second normalization, size-standardization, is predicated on the grounds that shape, in contrast to size, contains considerable informational content of biological importance. If size differences are at all appreciable, subtle shape differences may be swamped, and confounded. Three normalization approaches suggest themselves: [1] based on the A_0 term, [2] based on the arc length or perimeter of the outline and [3] based on the bounded area (van Otterloo, 1991; Lestrel, 1997). The first normalization refers to the constant of the FD series, A_0, which is defined as the mean of the vectors to the outline. Schwarcz and Shane (1969) were probably the earliest to apply this approach. Although this value is readily computed, and has been used as a scaling factor to control for size, it has the impediment that if the outline of interest departs significantly from a circle, the scaling factor will become increasingly inaccurate (Lestrel, 1980). The second normalization, based on the perimeter, has been applied in a pattern recognition context. A drawback with this normalization, for biological materials, is that significant differences can arise with a scaling factor based on the perimeter. For example, consider two similar forms with the exception that one of the outlines is relatively smooth and the other has extensive invaginations. The latter case will produce an unduly large perimeter in contrast to the former. The third approach, the one advocated here, is based on the area within the bounded outline. This approach is straightforward for planar single-valued outlines (Lestrel, 1974), but not for curves in 3-space (Section 9.5.5). An algorithm for the derivation of the area with conventional FDs can be found in Parnell and Lestrel (1977).

In spite of the extensive applications of the conventional FD to a wide variety of data sets, a number of constraints have limited the types of forms that can be numerically characterized using this approach. One of these problems is the presence of multi-valued functions. Multi-valued functions arise, for example, when outlines curve back on themselves (akin to retrograde motion). With respect to the cranial base (Lestrel and Roche, 1986), vectors drawn from the centroid to the vicinity of the posterior clivus will cross the morphology at more than one place on the outline. This results in a loss of data (Fig. 5 in Lestrel and Roche, 1986). Such situations are difficult to handle without casting part of the data into an imaginary plane, an unattractive solution (Lestrel, 1989a).

Another limitation with conventional FDs is the use of equally-spaced data. If homology of points is an issue, then it needs to be emphasized that it is not possible to incorporate homologous landmarks into a FD if equal angles are to be maintained. This makes such boundary methods 'landmark-free' with the exception of the landmark used for the initial orientation. While most conventional FDs tend to use equally-spaced points, this restriction can be overcome, but at the cost of considerably increased computational complexity because of the requirement of a weighted analysis.

Finally, recalling equation [9-8]:

$$f(x) = a_0 + \sum_{n=1}^{k} a_n \cos nx + \sum_{n=1}^{k} b_n \sin nx, \qquad [9\text{-}8]$$

an alternative, more compact discrete function can be generated by converting the sine and cosine terms in equation [9-8] to complex-exponential form by applying Euler's identity, $e^{\pm ix} = \cos x \pm i \sin x$. This allows the reduction of the sine and cosine terms to a single expression, c_n. Thus, equation [9-8] can be re-written as:

$$f(t) = \sum_{n=0}^{k-1} c_n e^{-i\,2\pi nt}. \qquad [9\text{-}16]$$

The two equations, [9-10] and [9-11], used to evaluate the Fourier coefficients, a_n and b_n, can also be converted in a similar manner, which yields:

$$c_n = \sum_{n=0}^{k-1} f(t) e^{i\,2\pi nt}. \qquad [9\text{-}17]$$

For both equations, [9-16] and [9-17], n is the usual harmonic number, t is time (viewed along the x-axis), e is the natural log base, and i is equal to $\sqrt{-1}$. The function variable, x, has been changed to a t here (the time axis), in keeping with the usual convention in signal analysis (see Tolstov (1962) for the derivations of equations [9-16] and [9-17]). We will return to these two important results in Section 9.6.

9.3.5. Applications Using FDs

The conventional FD has now been applied to a wide variety of data outside of engineering demonstrating its general utility as a function for fitting outlines.

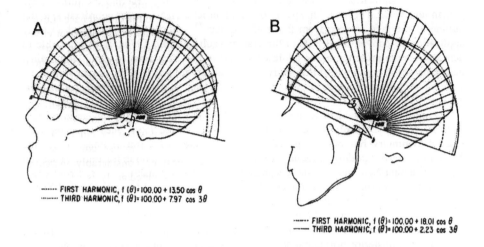

Figure 9.2. Cranial vault of *H. sapiens* and *H. neanderthalensis* (harmonics 1 and 3). Symbols: N, *Nasion*; eam, *external auditory meatus* (See text for explanation)

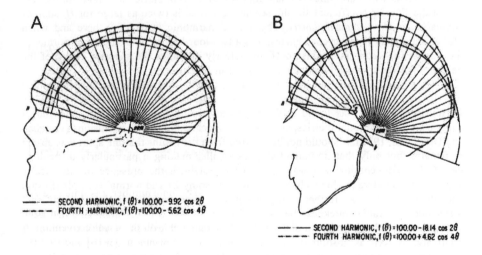

Figure 9.3. Cranial vault of *H. sapiens* and *H. neanderthalensis* (harmonics 2 and 4). Symbols: N, *Nasion*; eam, *external auditory meatus* (See text for explanation)

The earliest application of harmonic analysis to numerically describe the form of the human face was Lu (1965). That paper fitted a Fourier series, in polar coordinates, to the outline of the face as seen from the frontal aspect.[83] Nevertheless, it initiated an early application of conventional FDs to the cranial vault (Lestrel, 1974). It was possible to demonstrate the presence of the adolescent spurt in the cranial bones using longitudinal data. One of the outcomes of that study was the ability to clearly associate the numerical value of the harmonics with the visual information seen in the actual cranial form (Figs. 9.2 and 9.3).

Figure 9.2 illustrates a comparison of the cranial vault of: [A] *H. neanderthalensis* and [B] *H. sapiens* onto which harmonics one and three have been superimposed. Cranial symmetry has been imposed with the creation of a mirror-image about the *Nasion* to *external auditory meatus (n-eam)* line to generate a bounded outline (Section 9.3.4). The data have been partially standardized for size by making the A_0 term (which is equal to the mean of the vector lengths to the outline) equal to an arbitrary 100 and scaling the other coefficients accordingly.[84]

A look at the actual Fourier coefficients discloses that the first harmonic is significantly larger for *H. sapiens* compared to *H. neanderthalensis*. This coefficient measures the posterior horizontal shift of the external auditory meatus (*eam*) along the *n-eam* axis, which is greater in modern *H. sapiens*. The third harmonic coefficient, if graphed separately, has three lobes in polar coordinates and this 'shape' is remarkably close to the actual form of the *Neanderthal* cranium in contrast to *H. sapiens*.

Fig 9.3 is a comparison of the cranial vault between: [A] *H. neanderthalensis* and [B] *H. sapiens* with harmonics two and four superimposed. Particularly noteworthy here is that the Fourier coefficient for the second harmonic is twice as large for *H. sapiens* compared to *H. neanderthalensis*. The second harmonic is negative here and when plotted separately in polar coordinates, it is a two-lobed figure with the lobes normal to the *Nasion* (*n*)—*eam* axis. This coefficient clearly measures the superior extent of the height of the vault, which is considerably larger in *H. sapiens*.

The presence of the pubertal spurt was further investigated in a FD study, which was limited to the parietal bones in *Norma lateralis* (Lestrel and Brown, 1976). Curiously, the spurt was found in the males and not in the females. An independent study using conventional metrics, subsequently reached the same conclusions (Baughan and Demirjian, 1978). It should not be inferred from the results that the spurt was absent in females, but only that its magnitude was smaller making it particularly difficult to detect. The conventional FD was also used to establish the presence of statistically significant cranial vault shape changes between trisomy 21 and normal controls (Lestrel and Roche, 1976). A further study using the same sample displayed significant differences in cranial thickness between trisomy 21 and controls after size had been controlled for. Those results indicated that the cranial thickness in adult trisomics is relatively, as well as absolutely, thinner than normal controls, a consequence of the accumulating growth deficit associated with Down's syndrome (Lestrel and Roche, 1979).

---

[83] Unfortunately, the author was not able to replicate Lu's results.

[84] As indicated in Section 9.3.4, a more precise normalization for size is the area under the bounded outline.

A series of investigations were also carried out to numerically describe 2-D cranial base shape changes with age using both human (Lestrel and Roche, 1986) as well as non-human primate data (Lestrel and Sirianni, 1982). Systematic changes in both the human and in the *Macaca nemestrina* cranial base data could be discerned, the timing of which coincided with the pubertal spurt. All the above studies dealing with craniofacial data were based on FDs.

A number of papers using FDs have appeared in a forensic context. Inoue demonstrated that good discrimination between sexes could be obtained with just an analysis of the shape of the forehead (Inoue, 1990). This suggests the possible use of FDs with fragmentary skeletal materials. Another forensic approach where FDs could be potentially useful is in the matching of the skull with photographic sources (Pesce Delfino, *et al.*, 1986; Pesce Delfino, *et al.*, 1997). Problems remain with respect to orientation of the sources, skull and photograph. A recent study applied the conventional FD to assess the shape of the soft-tissue profile in *Norma lateralis* (Ferrario, *et al.*, 1992). A cluster analysis of the profile data produced four groups (two male and two female). However, it was not clear what criteria separated these four groups. Moreover, it was difficult to correlate the FD data with the more conventional cephalometric measures. Clearly, more work is needed before such FDs could be used for routine clinical use. Another paper applied the conventional FD to the shape of the mandible and found the Fourier coefficients correlated poorly with conventional cephalometrics (Halazonetis, *et al.*, 1991). An interesting application of FDs in neuroanatomy to characterize the shape of the brain using NMR imaging to establish cerebral asymmetries, can be found in Kennedy, *et al.*, (1990).

Extensive applications of the conventional FD to a wide variety of data are now available. These range from anatomy (Lestrel, *et al.*, 1977; Johnson, 1985; Johnson, 1997; O'Higgins and Williams, 1987; Le Minor, *et al.*, 1989) to cell biology (Ricco, *et al.*, 1985; Strojny, *et al.*, 1987; Murphy, *et al.*, 1990). However, constraints have limited the types of forms that can be numerically characterized using the conventional Fourier approach. Some of these have already been briefly mentioned, in particular, the presence of multi-valued functions. The analysis of both the primate and the human cranial base in *Norma lateralis* are such cases in point (Lestrel and Moore, 1978; Lestrel and Sirianni, 1982; Lestrel and Roche, 1986).

This Section is concluded with a brief discussion of the Zahn and Roskies algorithm, an independent FD, because of its relationship to eigenshape analysis (discussed in the next section). Rather than using a center as an origin from which equal radial measurements to the outline are taken, they define, instead, an angular function, f *(l), between segments on the closed planar outline (*i.e.*, where the outline is treated as a polygon) along its perimeter, L. They then expand f *(l) as a Fourier series over the interval 0 to 2π (Zahn and Roskies, 1972). According to Zahn and Roskies, the coefficients of their function are independent of starting point. The Zahn and Roskies Fourier descriptor contains all the information inherent in the curve outline allowing for the recreation of the form. Applications of the Zahn and Roskies algorithm include Waters, (1977); Ostrowski, *et al.* (1986); Kieler, *et al.* (1989).

9.4. EIGENSHAPE ANALYSIS

Another approach for numerically describing the outline, and thereby the shape of organisms, is eigenshape analysis. This technique was developed in a paleontological context and is based on the presumption that the use of eigenshape analysis facilitates the reduction of the morphological shape space to a comparatively few dimensions that contain most of the differences in shape (Lohmann, 1983; Schweitzer, *et al.,* 1986). Thus, it is claimed that eigenshape analysis provided a minimum number of factors necessary for recognizing subtle shape differences. Eigenshape analysis is based on a set of orthogonal shape factors (eigenfunctions) derived from a matrix of the correlations between shapes.

9.4.1. Algorithm for Eigenshape analysis

The basis for the eigenshape approach is the Zahn and Roskies (1972) algorithm. Rather than using a center as an origin from which radial measurements at equal angular intervals to the outline are taken, yielding, $r(\theta)$, they define an angular function, $\phi*(l)$, between segments on the closed 2-D outline (*i.e.,* the outline is treated as a polygon) along its perimeter, L, such that a polar expansion yields:

$$\phi*(l) = A_0 + \sum_{n=1}^{\infty} A_n \cos(nl - \alpha_n).$$ [9-18]

This equation includes the property of phase. The set $\{A_n, a_n; n = 1, \dots , \infty\}$ are the FDs for the curve. Note that for a circle $\phi*(l) = 0$. The values, A_n and a_n, are the amplitude and phase angle respectively for the nth harmonic (Zahn and Roskies, 1972). This formulation, $\phi*(l)$, is then expanded as a Fourier series over the interval 0 to 2π, which is invariant under translation, rotation and changes of the perimeter; that is, length or scale (Persoon and Fu, 1977).

Lohmann and Schweitzer (1990) proposed three measures for describing planar outlines within the context of eigenshape analysis. These were *form, size* and *angularity.* Form referred to two aspects of the outline: scaling (size) and amplitude (variance). Form was computed using the standardized (for size and variance) formulation of the 1972 Zahn and Roskies FD formulation. Size was defined two ways, as the perimeter of the outline and by the bounded area. As already mentioned, the use of perimeter cannot be recommended as a size measure because very convoluted outlines would have very large values while smooth outlines would have small values. Thus, area is the preferred variable for 2-D outlines (Section 9.3). Angularity was measured by the magnitude of the amplitudes of the angular part of the Zahn and Roskies algorithm; but, *before* these amplitudes were standardized to unit variance.

9.4.2. Procedures involved in Eigenshape analysis

Lohmann made use of the Zahn and Roskies procedure, equation [9-18], which computes Fourier coefficients using the tangent function, $f*(l)$. Lohmann was correct in indicating that if the centroid is not used, incorrectly computed amplitudes are the result.

This has been noted earlier by others (Parnell and Lestrel, 1977; Full and Ehrlich, 1982). Table 9.1 lists the steps required in applying the eigenshape procedure.

1. The 2-D outline is digitized generating a set of (*x, y*) coordinates of the periphery, which were measured at *n* equally-spaced points on the boundary.

2. The single-valued angular shape function (non-normalized) is then computed from the Zahn and Roskies algorithm. The perimeter is subdivided into an ordered set of angles between adjacent line segments to the equally-spaced points.

3. The angular shape function is then normalized with the removal of a circle of appropriate size yielding, $f*(l)$. This $f*(l)$ function can be thought of as measuring the departure of the outline from a circle. This step is intended to remove the effect of size.[85,86]

4. The amplitudes (as variance terms) are now derived from the shape function, $f*(l)$, and then further normalized to unit variance. This is a step prior to the computation of correlations between shapes.[87]

5. Homology is then preserved by rotating the function, $f*(l)$, so that a maximum correlation between forms is obtained. This is done using a single representative specimen as a reference shape.[88]

6. A matrix is then computed with the standardized angular values of the Zahn and Roskies function, $f*(l)$, (as columns) and specimens (as rows). Since the $f*(l)$ shape functions have now been normalized to zero mean and unit variance, it is possible to compute a new matrix of correlations. Using a similarity matrix based on pair-wise relationships between specimens, Lohmann then computed eigenvalues and eigenvectors. It is from these eigenvectors that the orthogonal eigenshapes are derived.

Table 9.1. Steps involved in eigenshape analysis.

Lohmann used this tangent function instead of the more conventional Fourier series in polar coordinates because of two advantages which the Zahn and Roskies FD

[85] This is equivalent to standardizing the conventional FD in polar coordinates with the use of the constant or, A_0 term (a circle), which is the mean of all the vectors (see Section 9.4).
[86] The greater the departure of the outline from a circle, the more the amplitudes of $f*(l)$ will increase in magnitude (a measure of angularity), and conversely, as the amplitudes of $f*(l)$ decrease to zero; the outline will converge to a circle.
[87] An argument has been made that this step may not always be desirable because it can make it difficult to distinguish between different shapes (Rohlf, 1986). Ray (1990) also indicated that this normalization was arbitrary and simply a consequence of the Zahn and Roskies' "removal of a circle" procedure and would not be necessary if conventional FD's were used.
[88] In effect a cross-correlation procedure exactly analogous to the one proposed earlier by Parnell and Lestrel (1977) using conventional FD's in polar coordinates.

presumably offers; namely, that it is always a single-valued function and that it does not require a vector center such as the centroid. However, with respect to the first issue, both the Zahn and Roskies formulation and the conventional Fourier series are largely limited to data consisting of relatively simple convex outlines. Morphologies with complex irregular multi-valued boundaries cannot be easily handled with either method.

Eigenvalues and eigenvectors have already been briefly encountered (Section 7.1.3). The principal eigenvector (the first eigenshape) represents the mean shape of the sample. The remaining orthogonal eigenshapes represent the partitioned variance about this mean shape (Full and Ehrlich, 1986). That is, they measure the deviation from the mean shape. So that while the first eigenshape visually depicts the average shape, the rest, as departures from the mean, would not resemble the form (Ray, 1990).

From the above correlation matrix, Lohmann arrives at his R-mode principal component analysis. According to Ray (1990), the technique of Lohmann and Schweitzer defines the first eigenshape as the sample centroid, rather than representing the first principal component of variation as generally presumed in principal component analysis. The first principal component of variation is the second eigenshape. However defined, the R-mode principal component scores derived here are used to reduce the dimensionality in the data, a stated goal of eigenshape analysis.

A number of objections can be raised with the use of eigenshape analysis, which merit some caution with its application to biological data. Eigenshape analysis is misleading in one sense. It actually represents the sequential use of two techniques rather than a unique single development. Eigenshape analysis is the initial application of Zahn and Roskies' algorithm, the results of which are subsequently subjected to a factor analytic method. While this approach is not without value, the reduction to a few eigenshapes may lead to the loss of potentially significant information if the 'higher' eigenshapes are excluded, simply because they only explain a moderate percentage of the variability in outline shape. A separate issue discussed earlier, is that the Zahn and Roskies algorithm is based on tangents, which can be difficult to measure on an outline. This makes it particularly sensitive to 'noise' and suffers from a closure problem. These problems became apparent in a comparison of a number of Fourier methods (Rohlf and Archie, 1984), and led to the conclusion that the Zahn and Roskies algorithm was the least satisfactory method (Rohlf, 1986).

It also needs to be noted that, the Zahn and Roskies representation is only valid for relatively simple planar outlines. It would be a poor choice for analysis of craniofacial structures such as the mandible or cranial base, which contain considerably more complexity than the comparatively smooth microfossil outlines analyzed by Lohmann and others. Moreover, the Lohmann (1983) contention that the Zahn and Roskies algorithm is preferred over conventional FDs is unwarranted and as Rohlf (1986) has demonstrated, it is equally possible to use conventional FDs as well as EFFs for the same purposes.

Finally, Lohmann's use of homology is rather perplexing if one defines homology as a one-to-one mapping between points or structures. To begin with, there are generally few available points that are homologous in a strict sense to allow for this mapping. Lohmann and Schweitzer (1990) have suggested that eigenshape analysis can be used to determine an 'average shape'. This is perhaps possible for simple smooth microfossils, but not for other more complex morphological structures. The ability to accurately handle homologous structures can become problematic (see Ray, 1990). While cross-correlation may be desirable in certain instances, it cannot insure homology. In fact, it can have the opposite effect. As alluded to earlier, the use of n equal-divisions particularly mitigates against the possibility of maintaining homology. In addition, as Full and Ehrlich (1986) have argued, there is very little possibility that homology between outlines can be consistently maintained with eigenshape analysis. Thus, Lohmann's approach is simply not robust enough or generally applicable to insure that homology can be consistently preserved (Full and Ehrlich, 1986; Ehrlich and Full, 1986). One suggestion that has been proposed for maintaining homology is to use unequal divisions by dividing the outline between homologous points into equal intervals, but maintaining the one-to-one mapping at the homologous points (Ray, 1990). However, as Ray indicated, this approach, while better at insuring homology, has other problems when applied within an eigenshape context. This issue of preserving the homologous point information within the boundary outline representation will reappear with the use of EFFs (see next section).

It is for these reasons that deformation models are more appropriate for maintaining homology, although at the expense of the boundary information. Without substantial further developments, eigenshape analysis, together with conventional Fourier analysis, must be considered as largely homology-free boundary methods. This is not to denigrate these methods, but rather to indicate that the strengths of these boundary methods lie elsewhere; specifically, in their ability to capture the global elements of form.

9.5. ELLIPTICAL FOURIER FUNCTIONS

The development of EFFs represents a comparatively new development, which circumvents a majority of the constraints that have characterized both conventional FDs as well as eigenshape analysis. It spurred FD developments in pattern recognition, for example, dealing with the rapid identification of military aircraft (Wallace and Mitchell, 1980; Kuhl and Giardina, 1982).

A computer program has been written to calculate various estimates using EFFs (see Appendix I). Both open and closed (bounded) curves can be handled with this specially written EFF software. Open curves are 'bounded', so to speak, by a procedure that requires digitizing from the first point to the last point along the boundary outline and then backtracking to the first point. However, it must be noted that this approach of handling open curves precludes normalization for size using the area since the area is effectively zero. An alternative normalization in this case would be the perimeter.

The requirement of equal intervals along the outline is now removed and the problem of multi-valued functions has been eliminated with this elliptic formulation. Computation is simpler since instead of the required integral solutions for the Fourier coefficients, the EFF coefficients are solved using an algebraic approach. Finally, the reduction of constraints associated with conventional FDs confers a decided advantage in that EFFs allow for the numerical characterization of a much larger class of 2-D shapes than heretofore possible.

9.5.1. The Kuhl and Giardina Parametric Formulas

The Kuhl and Giardina (1982) equations are a parametric solution set up in Cartesian coordinates in contrast to polar. That is, the formulation consists of a pair of equations (x and y) derived as functions of a third variable (t). Again, simplifying by letting the length of the period, L, be over a 2π interval, these parametric functions are defined as:

$$x(t) = A_0 + \sum_{n=1}^{k} a_n \cos nt + \sum_{n=1}^{k} b_n \sin nt, \qquad [9\text{-}19]$$

and

$$y(t) = C_0 + \sum_{n=1}^{k} c_n \cos nt + \sum_{n=1}^{k} d_n \sin nt, \qquad [9\text{-}20]$$

where n equals the harmonic number and k equals the maximum harmonic number. Kuhl and Giardina derived estimates of the Fourier coefficients that do not require integration. Once the expected x- and y-coordinates have been separately computed, they can be re-joined to produce orthogonal ellipses. The Fourier coefficients for the x-projection are:

$$a_n = \frac{1}{n^2 \pi} \sum_{p=1}^{q} \frac{\Delta x_p}{\Delta t_p} \left[\cos(nt_p) - \cos(nt_{p-1}) \right], \qquad [9\text{-}21]$$

and

$$b_n = \frac{1}{n^2 \pi} \sum_{p=1}^{q} \frac{\Delta x_p}{\Delta t_p} \left[\sin(nt_p) - \sin(nt_{p-1}) \right], \qquad [9\text{-}22]$$

where q is the total number of points along the polygon; as before n is the harmonic number. Here t_p is the distance between point p and point $p+1$ along the polygon, and x_p and y_p are the respective projections of the segment p to $p+1$. The Fourier coefficients for the y-projections are:

$$c_n = \frac{1}{n^2 \pi} \sum_{p=1}^{q} \frac{\Delta y_p}{\Delta t_p} \left[\cos(nt_p) - \cos(nt_{p-1}) \right], \qquad [9\text{-}23]$$

and

$$d_n = \frac{1}{n^2 \pi} \sum_{p=1}^{q} \frac{\Delta y_p}{\Delta t_p} \left[\sin(nt_p) - \sin(nt_{p-1}) \right]. \qquad [9\text{-}24]$$

Besides the four coefficients a_n, b_n, c_n and d_n that need to be evaluated, two constants A_0 and C_0 are also required. The constants B_0 and D_0 are equal to zero, which is analogous to conventional FDs. The constants A_0 and C_0 are computed from:

$$A_0 = \frac{1}{2\pi} \sum_{p=1}^{q} \frac{\Delta x_p}{2\Delta t_p} \left[t_p^2 - t_{p-1}^2 \right] + \alpha_p \left[t_p - t_{p-1} \right],$$ [9-25]

and

$$C_0 = \frac{1}{2\pi} \sum_{p=1}^{q} \frac{\Delta y_p}{2\Delta t_p} \left[t_p^2 - t_{p-1}^2 \right] + \beta_p \left[t_p - t_{p-1} \right].$$ [9-26]

The α_p and β_p terms needed above are:

$$\alpha_p = \sum_{j=1}^{p-1} \Delta x_j - \left[\frac{\Delta x_p}{\Delta t_p} \sum_{j=1}^{p-1} \Delta t_j \right],$$ [9-27]

and

$$\beta_p = \sum_{j=1}^{p-1} \Delta y_j - \left[\frac{\Delta y_p}{\Delta t_p} \sum_{j=1}^{p-1} \Delta t_j \right],$$ [9-28]

where $\alpha_1 = \beta_1 = 0$. (For details of the above formulations, see Kuhl and Giardina, 1982; Lestrel, 1989a; 1989b; 1997b).

If the boundary of a form can be modeled as *a curve in 3-space* (not as a volume), then the above equations can be extended without difficulty with the addition of a third parametric equation (Lestrel, 1997b). This parametric Fourier series in $z(t)$ is:

$$z(t) = E_0 + \sum_{n=1}^{k} e_n \cos nt + \sum_{n=1}^{k} f_n \sin nt,$$ [9-29]

The E_0, e_n and f_n coefficients being computed in an identical fashion to the 2-D case above (Lestrel, 1997b). An application of the 3-D EFF to characterize the boundary of the rabbit orbit can be found in Lestrel, *et al.* (1997).

9.5.2. Amplitude, Power and Phase Relationships

From the above equations, one can derive the usual values of amplitude, power and phase relationships. These variables are depicted in flow chart format in Appendix II. However, it is emphasized, that these parameters need to be computed separately for each of the coordinate axes (x, y, and z if a curve is to be analyzed in 3-space).

9.5.3. Other Elliptical Fourier Function Parameters

Since the above parametric formulations, for either the 2-D case or 3-D curve in space, produce ellipses, a number of additional measures that are potentially useful as 'similarity measures' for clustering and discriminant procedures, are available. These are ellipse area, perimeter, semi-major and minor axes, as well as angulation of the major ellipse axes with the x-axis or with respect to other ellipses. All these estimates again, are separately computed for each harmonic. These variables are also illustrated in flow chart format in Appendix II. Finally, it should be mentioned that a number of other estimates such as the area, perimeter, centroids and moments, have also been utilized (Kiryati and Maydan, 1989; van Otterloo, 1991).

9.5.4. Positional-orientation and Size-standardization Revisited

The two normalizations mentioned earlier (Section 9.3.4) positional-orientation and size-standardization apply equally well to EFFs. In the first instance, orientation, this is needed for superimposition purposes. Since Cartesian coordinates are involved in contrast to the polar, orientation is accomplished by a different procedure from that used with conventional FDs. An 'internal' orientation, as proposed by Kuhl and Giardina (1982), was utilized. This entails orienting each form by rotating it on the centroid until the major axis of the first ellipse is parallel to an axis, the usual case being the x-axis.

Kuhl and Giardina initially proposed the following procedure for size standardization. The value of the semi-major axis of the first harmonic ellipse was enlarged or reduced to 1.0 and the rest of the EFF coefficients scaled accordingly. For reasons alluded to earlier, this procedure is considered only partially successful because only one dimension is involved. In contrast, normalization for size was carried out in a similar fashion to conventional FDs. That is, the bounded area was calculated and scaled up or down to a constant $10,000 \text{ mm}^2$. The ratio of 100 divided by the square root of the original or actual area is then used as the scaling factor by which all EFF coefficients are subsequently multiplied. Application of these two procedures allows EFFs to be considered as having an orthonormal basis.[89]

9.5.5. Homology Once More

One of the early criticisms of conventional FDs has been that the homology of points across forms was lost, and with it, the localization of boundary features. With EFFs, this problem has been partially rectified. Since the EFF must be a close analog of the observed form, differences between the observed points and the actual curve being fractions of millimeters, it is possible to use the function to compute a set of predicted x, y and z coordinate estimates from the observed values as a function of t. This approach can be visualized as if the original observed homologous points are 'moved' (in technical terms, mapped onto the function) from observed space to the EFF function space. Homology is now maintained by a specific computational procedure: The first homologous predicted point is computed to be at the same location on the Fourier approximating (interpolating) function as the first observed point is on the digitized form.

---

[89] An orthonormal basis implies that the coefficients of a function are both orthogonal and normalized.

The second and subsequent predicted points are computed so that they have the same arc length from the first computed point as their counterpart (observed) landmarks do from the first digitized point on the original, digitized curve. This is equivalent to moving or translating the observed co-ordinates of the polygonal representation of the form, onto the EFF curve. This in effect maps the pseudo-homologous points from the digitized curve onto the EFF (Wolfe, 1997).

These points have been termed pseudo-homologous—after the suggestions of Sneath and Sokal (1973). It is incumbent upon the investigator to keep the residual, or difference between the observed points and the predicted points derived from the EFF, as small as practically possible. Mean residual values based on all points should not rise above 0.1 percent. That is, these values need to be well below the errors arising from such tasks as: [1] locating the points, and [2] digitizing them. Thus, the EFF now maintains the homology of the points, and with it, the homology of the entire form. While some loss is inevitable in regions of sharp curvature, this is usually minor and quite localized. Point homology can now be maintained, with the caveat that a sufficient number of homologous or pseudo-homologous points must be initially available. This approach insures the generation of precise *mean* morphological forms as well as allowing for a more precise statistical analysis of form size and shape differences. Thus, this technique can be considered as the first tentative step toward a model that begins the integration of the homologous point data with the boundary outline information (Lestrel, 1997b; Lestrel, *et al.*, 1997).

Nevertheless, the Fourier analytic methods above, like the landmark methods earlier, are not without constraints. Because the contributions of each of the amplitude coefficients $(a_1, a_2, \cdots, a_n; b_1, b_2, \cdots, b_n)$ are global in nature, the localization of a particular frequency in the time domain, or a boundary discontinuity in the spatial domain, cannot be easily determined. In other words, localized features on the boundary are 'smeared' over the total form and cannot be 'locked on', so to speak, with the amplitudes. To identify where these localized boundary aspects occur, alternative methods have been required, which have tended to be, up to now, subjective. One approach to identify these localized aspects has been to compute specific distances from the centroid to homologous boundary points of interest (Lestrel, 1997b and elsewhere).

Conventional Fourier Analysis	Elliptical Fourier Functions
Data sets in polar coordinates (r, θ)	Data in Cartesian coordinates (x, y)
Requires equally-spaced intervals	Not limited to equal intervals
Limited to single-valued forms	Not limited to single-valued forms
Limited computational complexity	Greater computational complexity
Requires evaluation of integrals	Algebraic solutions
Point-homology lost	Point-homology can be retained
Possible to relate numerical differences in the coefficients with differences in the observed shape using reflection (mirror-imaging)	No readily apparent relationship between numerical differences in the coefficients and differences in the observed form

Table 9.2. Advantages and limitations of Fourier descriptors.

It is this lack of specificity with respect to localization of boundary elements that has spurred the development of wavelets (Section 9.7). Although the emphasis in the next chapter is on structural morphometrics per-se, wavelets also now play an increasing role in shape analysis with respect to the boundary. Table 9.2 lists some of the advantages and limitations of conventional FDs and EFFs as applied to bounded morphological forms.

9.5.6. Applications Using EFFs

A number of papers have appeared since Kuhl and Giardina published their EFF algorithm in 1982. While the emphasis was initially directed toward pattern recognition (Lin and Hwang, 1987), two publications, one dealing with the shape of mosquito wings (Rohlf and Archie, 1984), and the other on the outline of *Mytilus edulis* shells (Ferson, *et al.*, 1985) represent the initial extension of EFFs in a zoological context.

Rohlf and Archie (1984) compared the shape of mosquito wings using EFFs, the Zahn and Roskies formulation and conventional FDs. They found that EFFs produced the most satisfactory results in terms of discrimination. A series of studies in a botanical framework compared the shape of plant leaves using conventional measures (CMA), Freeman chain codes, moment invariants as well as EFFs (White and Prentice, 1988; White, *et al.*, 1988). They also found that EFFs yielded the best discrimination between groups. A study of the primate cranial base in *Norma lateralis* using EFFs standardized for size, yielded equivalent results with respect to discrimination (Lestrel, *et al.*, 1988). These studies were briefly mentioned in Section 7.3.2. Using EFF amplitudes for the four extant adult groups: [1] *H. sapiens* (n=31); [2] *P. troglodytes* (n=22); [3] *G. gorilla* (n=10); and [4] *M. nemestrina* (n=29) produced an almost perfect classification using discriminant functions. That is, out of 92 specimens, only one chimpanzee (*P. troglodytes*) was misclassified as a gorilla. A later study added a fifth group of baboons, *P. cynocephalus* (n=30). Utilizing a different set of variables, centroid-based distances and first ellipse parameters, a 100% discrimination was subsequently obtained (Lestrel and Swindler, 1996).

Recent investigations of both the cell boundary and the nucleus of human leukocytes utilizing EFFs produced characteristic differences in shape for the classes: neutrophils, lymphocytes and monocytes (Diaz, *et al.*, 1989; Diaz, *et al.*, 1990; Diaz, *et al.*, 1997; Nafe, *et al.*, 1992). Satisfactory convergence onto the cell outline with the EFF was already apparent after only four harmonics. Differences could be demonstrated using the invariant properties of the separate ellipses produced by the EFF, such as their area and the semi-major and semi-minor diameters, etc. Moreover, these geometric properties based on the first harmonic were sufficient to discriminate these cell types. However, discrimination of similar cell types such as normal blood lymphocytes and chronic lymphocytic leukemia cells necessarily required the evaluation of the higher harmonics. Subsequent applications dealing with craniofacial structures such as the maxilla, mandible, nasal bones, facial profile and mastication have also appeared (Lestrel, 1987; Lestrel, *et al.*, 1991; Ferrario, *et al.*, 1990; Lestrel, 1997b). One of these structures, the mandible, is illustrated in Figure 9.4. The mandible is shown here with the EFF as a stepwise fit.

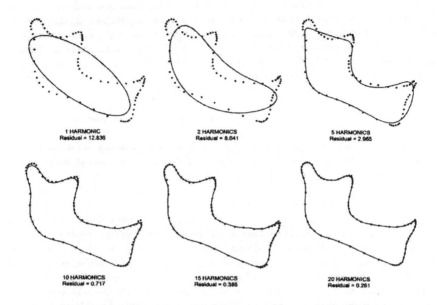

Figure 9.4. Stepwise fit of the EFF to the mandibular morphology.
With 20 harmonics, the mean residual[90] has fallen to 0.26 mm

A study of mandibular response to functional appliance therapy provided a particularly instructive example of the usefulness of EFFs to characterize morphological shape changes in a clinical setting. The effects of functional appliance therapy (applied to Class II malocclusions here) have been, at times, controversial leading to conflicting findings. Prior use of CMA had been inconclusive in resolving the discrepancies seen in the various studies. The application of EFFs to numerically describe the mandible in global terms revealed that the treatment (*Tx*) effect resulted in a significant posterior condylar bending; a bending that could be described in numerical as well as visual terms.

This bending explained the observed increase in mandibular length that some clinical investigators had attributed to increased growth, supposedly stimulated with functional appliance therapy (Moon, 1992; Moon, *et al.,* 1992; Moon, 1997). Another study using EFFs assessed the *Tx* effects of functional appliance therapy in Class III's, also demonstrated statistically significant changes. In this case, *Tx* resulted in a posterior repositioning of both the maxillary and mandibular incisal margins. The *Tx* effects where more locally limited to the dentoalveolar aspect and did not involve the basal areas of either the maxilla or mandible (Lestrel and Kerr, 1992; 1993).

A more recent study applied EFFs to characterize the shape of the cranial base in *Macaca nemestrina* (Lestrel, *et al.,* 1993). This study represents an extension of

---

[90] As indicated earlier, the residual is computed as the difference between the observed points and the predicted points derived from the EFF. The residual values are then averaged over the whole form.

previous work using conventional FDs (Lestrel and Moore, 1978; Lestrel and Sirianni, 1982). Statistically significant changes were found for both sex and age. The age changes consisted of a gradual lengthening in the anteroposterior direction with a simultaneous narrowing in the supero-inferior direction. An elongation of the *dorsal clivus* as well as an anterior migration of the *hypophyseal fossa* was observed. Work has also focused on precisely delineating the location of 2-D shape changes in the lateral view of the cranial base in shunt-treated hydrocephalics compared to normal age and sex matched controls (Lestrel, *et al.*, 1994; Lestrel and Huggare, 1997). Thus, FDs and especially EFFs can be considered as useful methods for documenting, both numerically as well as visually, global shape changes in complex 2-D morphologies of the type encountered in the craniofacial complex.

The extension to 3-D alluded to earlier, represent a more recent development of the application of EFFs to data. This extension facilitates the numerical description of outlines that can be modeled as a curve in 3-D (not as a volume). As noted earlier, the area is needed for size-standardization and the centroid is required for superimposition and comparison of mean forms. These two properties would seem to be straightforward extensions of the 2-D case. Unfortunately, this is not the case for a closed curve in 3-D.[91] Here, the area within a bounded outline in 3-D (as a curve in space), needs to be viewed as the 'hole' in a donut with the donut body having almost no 'thickness'. Such a structure can be treated as 'minimum surface'. The concept of a minimum surface has held considerable fascination for mathematicians and laymen alike. In the absence of computers, and until quite recently, the computation of such minimal surface areas represented a major challenge. D'Arcy Thompson, for example, considered the form of the radiolarian skeleton as a minimum surface controlled by surface tension (Almgren, 1982).

Soap films are the most common examples of such minimum surfaces (Almgren and Taylor, 1976). One can view the soap film as a surface that is seeking to minimize surface tension. This minimization of tension can also be modeled as minimizing energy. The area and centroid location of such boundary outlines are difficult if not impossible to adequately describe in conventional numerical terms. However, by utilizing an energy minimization approach, it is possible to circumvent these difficulties. This technique is developed in detail elsewhere (Brakke, 1992; Lestrel, *et al.*, 1997).

The morphology in question was the rabbit orbit, used to model the growth of the eye and its surrounding structures (Sarnat and Shanedling, 1972; Sarnat 1981). The bony margin of the orbit can only be effectively visualized as a bounded curve in 3-D, hence the need for a 3-D EFF curve fit. A sample of 44 rabbit skulls, infants, juveniles and adults were available to access form changes due to growth. These were measured with a set of 12 distances from the 3-D centroid to selected aspects on the orbital outline. Figure 9.5 shows a 3-D plot of the mean orbit outlines of the infant (n=14), juvenile (n=9) and adult (n=21) samples superimposed on the 3-D centroid. Pronounced size changes can be observed. The data shown are *not* normalized for size. However, by standardizing for size, using the minimum surface area discussed above, and superimposing on the centroid, shape changes from infant to adult were also plotted (not

---

[91] As mentioned earlier, open curves, either in 2-D or in 3-D, do not allow for size standardization using area. In this case, the perimeter can be considered as a suitable normalization.

shown). Statistical results indicated that significant shape changes occurred with growth. These shape changes were particularly pronounced at the *supraorbital* process (Okamoto, 1994; Osman, 1993; Lestrel, *et al.*, 1997).

RABBIT ORBIT - SIZE AND SHAPE

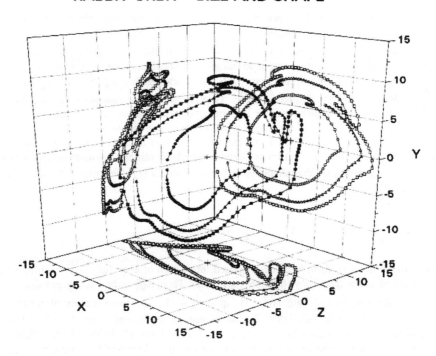

Figure 9.5. The Bounded outlines of the rabbit orbit in 3-D. Symbols: infants (filled circles); juveniles (filled triangles); and adults (filled squares). Projections shown are in the lateral (X), superior (Y) and A-P (Z) planes. Superimposition is on the 3-D centroid.

9.6. FOURIER TRANSFORMS

9.6.1 The Discrete Fourier Transform

The relationship between FDs and the Fourier transform (FT) will be briefly examined here. The continuous FT, as well as its cousin the DFT, is a mapping from the time or spatial (boundary) domain into the frequency domain, in an exactly analogous way to FDs. The FD, as well as the FT and DFT, represent a decomposition of the waveform into different frequency sinusoids which, when combined, can be used to re-create the waveform used as a representation of the original function, $f(t)$. The continuous FT represents a more general case in that it can be used to analyze non-periodic phenomena or non-stationary signals in contrast to FDs, which require that the data be periodic. Thus, the FD is simply a special case of the FT or DFT (Brigham, 1974; Challis and Kitney, 1991).

Non-periodic waveforms are those that do not repeat themselves. Consider a periodic function where each harmonic is separated by an amount $\Delta f = 1/L$, where Δf is the change in frequency and L is the period. In brief, the FT is derived by taking the discrete Fourier series, converting it to complex-exponential form and then taking the limit so that as the change in frequency, Δf, approaches zero the period, L, approaches infinity (Ramirez, 1985; Brigham, 1974). Thus, a periodic or non-periodic waveform, given by *f(t)*, can be transformed or decomposed from a function over time, *f(t)*, (or space when dealing with form) into a new function, containing sinusoidal frequency components or amplitudes. The FT is composed of two parts (called a Fourier transform pair). The first part is the Fourier transform, which identifies the different frequency components (amplitudes) that characterize the function, $f(t)$. This new function, *X (f)*, is defined as:

$$X(f) = \int_{-\infty}^{\infty} x(t)e^{-i2\pi ft}dt , \qquad [9\text{-}30]$$

where *x(t)* is the waveform to be decomposed into a sum of amplitude coefficients, *X(f)*, as a function of the frequency (harmonic analysis). Thus, *X(f)* is the Fourier transform of *x(t)*, *e* is the natural log base and *i* is equal to $\sqrt{-1}$, and the exponential function, $e^{-i2\pi ft}$, is called the kernel. These Fourier amplitude coefficients contain both real and imaginary sinusoidal components. The second part of the FT is the inverse Fourier transform (harmonic synthesis), which allows one to 'transform back' from the frequency domain function, *X(f)*, to reconstruct the actual signal or spatial form, *x(t)*. In other words, to recreate the non-periodic waveform, *x(t)*, from its Fourier transform:

$$x(t) = \int_{-\infty}^{\infty} X(f)e^{i2\pi ft}df , \qquad [9\text{-}31]$$

In contrast, the DFT is the algorithm of choice with actual data if the latter is sampled as a set of discrete points, generally evenly distributed in either time or space. Additionally, because the FT involves a mapping into the complex plane, it, presumably, can handle multi-valued functions similarly to EFFs (Thomas, *et al.*, 1995). Consider a complex plane, *z(n)*, in which the coordinates of a point on the 2-D outline, composed of *N* points, can be defined as:

$$z(n) = x(n) + iy(n).$$ [9-32]

Then the common form of the DFT in exponential notation is:

$$F(k) = \frac{1}{N} \sum_{n=0}^{N-1} z(n) e^{-i2\pi kn/N},$$ [9-33]

where *F(k)* is the discrete form of the continuous FT, equation [9-31]. *N* is the total number points on the form, *1/N* is a scaling term. The inverse DFT is given by:

$$z(n) = \sum_{k=0}^{N-1} F(k) e^{i2\pi kn/N}.$$ [9-34]

Note the similarity of equations [9-33] and [9-34] with equations [9-16] and [9-17]. Alternatively, equations [9-33] and [9-34] can be re-expressed in terms of sine and cosines Again, changing notation and simplifying by setting the period equal to a 2π interval, gives the DFT as:

$$X(t) = \frac{1}{k} \sum_{n=0}^{k-1} x(n) \cos nt - ix(n) \sin nt,$$ [9-35]

and the inverse DFT as:

$$x(n) = \sum_{t=0}^{k-1} X(t) \cos nt + iX(t) \sin nt,$$ [9-36]

where the cosine term is the real *n*th component of the DFT and the sine term refers to the complex *n*th component (modified from Ramirez, 1985). While these equations are not generally used for computational purposes in contrast to equations [9-33] and [9-34], they facilitate comparison with conventional FDs; that is, equation [9-8].

A common application of the FT has been to transform digitally recorded music into the complex Fourier domain and filter out unwanted elements, and then re-convert the changed transform back into an improved musical signal. Developments such as the FFT and other formulations such as the Hartley transform are particularly useful here (Bracewell, 1989).

While the FT approach is often used to numerically describe textural characteristics using, for example, optical data analysis (Chapter 10), it has been also used to characterize outlines in an analogous way to conventional FDs. The FT was applied to characterize the shape of leaves (Kincaid and Schneider, 1983). Fourier coefficients were generated from the leaf periphery. As these coefficients contain both real and imaginary components, both were utilized. As the number of harmonics approach $2^{N/2}$, the approximation converges onto the leaf outline. This is exactly analogous to the FDs discussed earlier. Area was normalized by dividing both the real and the imaginary component, by the square root of the leaf surface area. This resulted in differences in leaf form based on shape only. Statistical tests of mean differences in leaf shape substantiated the visual impression, that dissimilar leaf shapes also reflected significant differences in the amplitudes.

Another interesting application of the FT is the description of the internal and external patient contours as seen on a CAT scan image. The intent of these authors was to develop a method that would allow the storage and retrieval of this contour data (Mok and Boyer, 1986). The number of points on the periphery was a power of two to satisfy FFT requirements (Section 9.6.2). As expected, most of the contour information was contained in the lower frequency components, so the FT was truncated at 50 harmonics and the higher frequencies removed. The inverse transform was then applied and the boundary outline recreated which closely approximated the original contour.

All these Fourier methods, whether derived from either boundary or structural morphometric considerations (since there is considerable overlap) are, nevertheless, still subject to limitations. Specifically, all the Fourier methods described above, with the exception of EFFs, contain two restrictions, which are: [1] that equal intervals are generally required for the time/boundary representation and [2] that the points, n, must be sampled as a power of two (2^n) if the FFT is to be utilized (see next section). These restrictions remain constraints that may cause difficulties with data where point homology plays an important consideration (Bookstein, 1991).

9.6.2. The Fast Fourier Transform

Finally, the DFT represents computational advantages in conjunction with the development known as the fast Fourier transform (FFT). The FFT cuts down the computer time required to calculate the transform coefficients as the number of harmonics become large (Ramirez, 1985; Hamming, 1973; Bracewell, 1989). Whether one is using conventional FDs or FTs, computer time can become problematic. In terms of actual calculations using the FT, one has to do n additions plus n^2 multiplications (Bracewell, 1989). For example, consider a curve containing 1024 points on its periphery (the choice of this particular number will be made clearer subsequently); the number of multiplications would be 1024x1024 or 1,048,576 plus the 1024 additions, requiring over one million computations! This will slow down even machines with comparably fast CPUs. To circumvent these increasingly long CPU times, Cooley and Tukey (1965) developed an alternative algorithm, which starts with the boundary outline being divided into a large number of equal intervals, with the restriction that these numbers must be equal to 2^n, where n is the point number.

If one were to then divide the total number of points (say, 1024) into two halves with each half containing one-half the number of points (512, 512) then a savings in the number of computations can be achieved. Thus, the reduction from 1024 to 512 results in a total of 262,656 multiplications and additions for one 'piece' of the curve, adding together the two halves (262,656 + 262,656 = 525,312), results in a savings of 50%. Continuing in this vein, if one were to do continuing subdivisions until only pieces containing two points were left, would that substantially decrease the number of computations required? The Cooley and Tukey (1965) algorithm clearly answers this in the affirmative.[92]

We can now re-construct the whole curve in steps by a process of combining 512 of the two-point pieces with 256 of the four-point pieces, then with 128 of the eight-point pieces and so on, until one arrives back at the 1024 point outline. Ignoring additions because they are negligible, this procedure is highly efficient in terms of multiplications since in this example there are now only ten steps involved ($1024 = 2^{10}$), each of which requires 1024 multiplications. This generates 10x1024 or 10240 multiplications instead of the 1,048,576, which is a saving of over 99%! Thus, these CPU times, initially N^2 where N is the number of points are now reduced to $N * \log_2 N$ calculations with the FFT (Cooley and Tukey, 1965; Brigham, 1974; Newland, 1993).

9.6.3. The Short Time Fourier Transform

As mentioned earlier, the inability to precisely identify position in the time/boundary representation has spawned new methods. The discrete wavelet transform (DWT) represents one such response to this inability of earlier Fourier methods to identify localized boundary phenomena. Prior to the DWT however, an earlier approach was developed in an attempt to deal with this lack of localization; in effect, a revision of the FT. This was the windowed or short time Fourier transform (STFT) attributed to Gabor, who utilized a window of finite length that was shifted, starting at *t=0* in the time domain (or *n=0* in the spatial domain) along the time/boundary representation. Here the signal, *z(n)*, is now multiplied by the window function, *w(n)*, such that:

$$F^{STFT}_{(n', k)} = \frac{1}{N} \sum_{n=0}^{N-1} z(n) w^*(n - n') e^{-i2\pi kn/N}, \qquad [9\text{-}37]$$

where the window is defined by $w^*(n - n')$ with the * representing the complex conjugate. The *n-n'* refers to the width of the window. This approach begins to identify the location of the frequency components in the time/boundary domain (Hubbard, 1996; 1998). However, while the STFT predates wavelets, it is based on a window function that has a *constant* width for all frequencies. The width of the window function is known as the *support*. If the window is narrow, it is called *compactly supported*. Good compact support also implies that the window function is zero on either side of the window interval. These concepts will arise again in discussions dealing with wavelets. The current wavelet transform (developed below) uses a multiresolution method allowing for a variable 'window' shift instead of the constant window characteristic of the STFT. In other words, it allows for the recognition of high frequency wavelets, which are narrow in width as well as low frequency wavelets, which are wide. This facilitates a more

---

[92] It is of some interest that the computational approach leading to the FFT was anticipated by the great mathematician Karl Gauss at the beginning of the 19th century, (Bracewell, 1989).

precise localization in both the time/boundary and the frequency domains, making wavelets superior to the STFT.

9.7. WAVELET ANALYSIS

In an analogous way to the FT, there are also continuous and discrete forms of wavelets. Additionally, as there are two sets of orthogonal functions (*i.e.,* sine and cosines) that determine the Fourier expansions and transforms, there are also two sets of orthogonal functions on which wavelet expansions and transforms are based (Nadler and Smith, 1993). The major difference between FTs and wavelets rests on periodicity. That is, FTs range over the period—∞ to +∞ for the continuous case, while wavelets are of limited duration and rapidly tend to zero at both ends of the interval for both the continuous and discrete cases. This results in a 'small' wave or wavelet. Thus, because of their finite duration they are narrowly focused in either time or space. Accordingly, this allows for the identification of localized information of the frequency components in the time/boundary representation, a property missing with conventional FTs.

9.7.1. The Continuous Wavelet Transform

In keeping with earlier FD considerations, the decomposition into the number of wavelet components or levels is determined by the number of points in the time/boundary domain, given the two constraints, equal intervals, and the number of points, n, subject to the 2^n restriction. This is analogous to FFT requirements. Thus, if the outline is sampled with 64 points, six levels are available ($2^6=64$), with 2048 points eleven levels are possible ($2^{11}=2048$). Therefore, the greater the number of points on the bounded outline, the higher the level, and the greater the detail in the outline (higher frequencies) that becomes available. Each of these levels represents a different resolution or scale. This scale is also termed an *octave*, which is derived from music since conveniently, each higher octave implies a doubling of the frequency (Strang, 1994). This division of the signal/boundary domain into levels is called a *multiresolution* decomposition (Mallat, 1989).

The shapes of each of the wavelet components (windows) depend on what is called an *analyzing wavelet* or *scaling function*. Once determined, the analyzing wavelet forms the basis functions into which the form under consideration will be decomposed. This analyzing wavelet has to meet certain requirements such as orthogonality of the wavelets with each other (Newland, 1993). Presented here is a rather intuitive, or symbolic, approach based on the continuous case of wavelet transforms. The continuous wavelet transform (CWT) can be set up as:

$$C(j,k) = \int_{-\infty}^{\infty} f(t)\Psi(j,k,t)dt , \qquad [9-38]$$

where j refers to a scaling factor and k to a position factor, $f(t)$ is the function to be decomposed into wavelets, and the term $\Psi(j,k,t)$ symbolically refers to the basis function of the wavelet. Therefore, from equation [9-38], the wavelet amplitude coefficients, $C(j,k)$, are a now a function of *both* scale and position.

This process involves choosing a *mother* wavelet from which one creates copies or *daughter* wavelets. Each of these copies are translations and dilations by powers of two of the mother wavelet. Specifically, one places the wavelet at the start of the function, $f(t)$, to be decomposed and computes $C(1,1)$, an amplitude coefficient that determines how closely the wavelet is correlated with the first signal/boundary section. One then shifts k-positions to the right and computes $C(1,k)$, coefficients until one reaches the end of the signal/boundary. One then returns to the beginning of the function, $f(t)$, but now with the wavelet *scaled*, either dilated or compressed, and computes a new scaled $C(j,1)$ coefficient, shifting to the right for k-positions and continuing the process of computing $C(j,k)$ amplitude coefficients for each increasing j-scale. This process then leads to a multiresolution representation.

While a large number of such analyzing wavelets are possible, they must satisfy certain constraints to be useful, so only a few have been widely utilized. Moreover, some wavelets are more suited than others depending on the application. Wavelets that have been utilized include the Haar wavelet originally devised in 1910, as well as numerous others such as the Gabor, Daubechies D4, Morlet, Mexican hat, Coiflet, etc., (Strang, 1994; Walker, 1999).

9.7.2. The Discrete Wavelet Transform

In the last decade, wavelet basis functions (in contrast to trigonometric basis functions seen with FDs, which are composed of sines and cosines) have been increasingly utilized. The essential difference between FTs and wavelets as already mentioned, is that while the former are periodic over the $-\infty$ to $+\infty$ interval, the later are composed of a finite interval that tends to zero on either side of the interval. Hence, wavelet, meaning little waves, refers to a (window) function of finite length or *compactly supported*. The wavelet transform represents an alternative decomposition of the signal/boundary domain into wavelet components or levels (multiresolution). Levels refer to the *scale* (to be defined subsequently), which now take the place of frequency in the STFT. When the separate wavelet levels are summed, the original form is re-constructed (information-preserving). The boundary is generally sampled at equal intervals and the n sampled points set up as a power of two (2^n). This approach is analogous to earlier Fourier methods with the exception that the basis functions are now different. Thus, while wavelets have their origins grounded in FD methods, they represent an independent development, as well as being necessarily more complicated. The following discussion will be limited to 1-D wavelets; 2-D wavelets will be taken up in Chapter 10.

While the CWT provides exact solutions, it makes considerable CPU demands; these however, can be obviated with the use of the DWT. The specific DWT to be described is called a harmonic wavelet transform (HWT) , which was developed by Newland (1993; 1994). It is depicted here because of some of its computational advantages. Recalling the DFT as:

$$F(k) = \frac{1}{N}\sum_{n=0}^{N-1} z(n)e^{-i2\pi kn/N}, \qquad [9\text{-}33]$$

and its inverse DFT as:

$$z(n) = \sum_{k=0}^{N-1} F(k)e^{i2\pi kn/N}, \qquad [9\text{-}34]$$

one can now utilize the harmonically-based DWT (HWT) to generate the localization in the time/boundary domain, $z(n)$, which is missing with the DFT. The HWT assumes that the function $z(n)$ is periodic. This circular wavelet function as set up as follows. Let the 'mother' wavelet, $W(x)$, describing the translations and dilations, take the form of:

$$W_{j,k}(x) = w(2^j x - k), \qquad [9\text{-}39]$$

with integers j and k, where scale is defined by j and position by $x=k/2^j$. This is assumed to be a circular wavelet in that it repeats over the interval, L, in an identical fashion to FDs. The expansion of $z(n)$ can now be approximated as a wavelet:

$$z(n) = a_0 + \sum_{j=0}^{\infty} \sum_{k=0}^{2^j-1} a_{2^j+k} w(2^j n - k), \qquad [9\text{-}40]$$

where there is now a double summation, since scale, j, and position, k, are involved. Equation [9-39] assumes that the function, $z(n)$, is real with real amplitudes a_{2^j+k}. If $z(n)$ is complex, then complex amplitudes are involved in an analogous way to the DFT and its inverse. That is, if $z(n)$ is real, $\tilde{a}_{j,k}$ is the complex conjugate of $a_{j,k}$ so that $\tilde{a}_{j,k} = a^*_{j,k}$; however, when $z(n)$ is complex, the expansion takes a somewhat different form:

$$z(n) = a_0 + \sum_{j=0}^{\infty} \sum_{k=0}^{2^j-1} \{a_{2^j+k} w(2^j n - k) + \tilde{a}_{2^j+k} w^*(2^j n - k)\}. \qquad [9\text{-}41]$$

Thus, since the HWT is defined in the complex plane, two amplitude coefficients, a_{2^j+k} and $\tilde{a}_{2^j+k}$, are now required (see Newland, 1993; 1994). This is analogous to the one coefficient, $C(j,k)$, in equation [9-31]. The amplitude coefficients in equation [9-41] are evaluated from:

$$a_0 = F(0), \qquad\qquad [9\text{-}42]$$

$$a_{2^j+k} = \frac{2}{k} \sum_{s=0}^{2^j-1} F_{2^j+s}\, e^{i2\pi sk/2^j}, \qquad\qquad [9\text{-}43]$$

$$\tilde{a}_{2^j+k} = \sum_{s=0}^{2^j-1} F_{N-(2^j+s)}\, e^{-i2\pi sk/2^j}, \qquad\qquad [9\text{-}44]$$

and

$$a_{N/2} = F(N/2) \qquad\qquad [9\text{-}45]$$

where $0 \le k \le 2^j - 1$ refers to the position on the boundary outline and $0 \le j \le J - 2$ refers to the limits of the scaling, s, or resolution level. Two FFT calculations are required. Amplitude coefficients a_{2^j+k} are obtained from the inverse FFT of the sequence $F(2^j+s)$. Amplitude coefficients $\tilde{a}_{2^j+k}$ are generated from the direct FFT of the sequence $F(N-2^j-s)$. Coefficient a_0 and $a_{N/2}$ represent special cases. Specifically, the a_0 term represents the mean of the sample, in an analogous way to conventional FDs, equation [9-9], while the $a_{N/2}$ term is an additional term reflecting the Nyquist frequency restriction (Drolon, *et al.*, 1999a; Drolon, *et al.*, 1999b). For details, the reader is directed to works by Newland and Drolon.

9.7.3. One-Dimensional Wavelet Applications

Applications of 1-D wavelets include signal processing, the de-noising of data, EEG diagnostics, speech synthesis, compression, etc. These are generally viewed in 1-D terms (2-D applications will be taken up in Chapter 10).

An area where wavelets have been applied has been in the recognition of hand-printed characters, the feature extraction of interest being the contour of the letters (Wunsch and Laine, 1995). As indicated earlier (Section 9.3.4), low low-frequency components deal with the global aspect of the boundary. With respect to handwriting, it turns out that this lower frequency information is less sensitive to the idiosyncrasies of handwritten characters, and thus, potentially useful for pattern recognition analysis. However, since a drawback with conventional FDs is that the localized information present is lost, wavelets are preferable here since they supply that information. Wunsch and Laine were able to ignore the high-frequency components and derive wavelet descriptors that provided reliable recognition of handwritten characters containing large variability.

Wavelets are also being used in sedimentology to characterize particle shape. This represents a continuation of research that started with the application of FDs and FTs (Schwarcz, and Shane, 1969; Ehrlich and Weinberg, 1970; Drolon, 1998). However, few applications have appeared in a biological context. One deals with human growth (Fujii and Matsuura, 1999). This paper, while interesting from a curve-fitting perspective, does not deal with the analysis of form per-se. It is expected that this paucity of wavelet-based biological research will be rectified in the near future.

KEY POINTS OF THE CHAPTER

This chapter dealt with boundary methods. These included median axis techniques, eigenshape analysis, as well as Fourier analytic procedures. The later consisting of conventional Fourier descriptors (FDs) as well as elliptical Fourier functions (EFFs). From FDs, there was a logical progression to the Fourier transform (FT), which initially led to the fast Fourier transform (FFT) to cut down the calculation time requirements. Problems with FDs and FTs, such as the inability to identify position in the time/boundary representation spawned new approaches. These started with the short time Fourier transform (STFT) which eventually led to the continuous (CWT) and discrete wavelet transforms (DWT), of which one particular version, the harmonic wavelet transform (HWT), was briefly discussed.

CHECK YOUR UNDERSTANDING

1. What are some of the advantages of boundary methods over coordinate methods? What are some of the disadvantages?

2. Discuss the method of median axis analysis and as a project; apply it to some biological data.

3. If you use conventional FDs in polar coordinates and create a mirror image of the 2-D data, why will this configuration allow the attachment of biological meaning to the amplitudes? Hint: Carefully analyze Figs. 9.2 and 9.3.

4. Given exercise #3, can you think of any other biological structures that might profit from such an approach?

5. What is the relationship of the amplitude spectrum to the power spectrum? How is the concept of variance related to power?

6. What is phase? How can it play a role in the analysis of the form of biological organisms?

5. List the advantages and especially the constraints that are involved in the usage of the Zahn and Roskies' algorithm. How do these apply to eigenshape analysis?

6. Can the procedure of EFFs be considered superior to conventional FDs? If so, give some reasons and attempt to justify them.

7. What are some of the differences between conventional FDs and EFFs? Do these differences play a role in the type of organismal form to be analyzed?

8. Is it possible to attach meaning to the amplitudes using EFFs? How would you go about doing it?

9. How does homology play a role in the use of FDs? How can the issue of homology be handled with the utilization of EFFs?

10. How are EFFs used to analyze data in 3-D? Can you think of some examples that are suitable for this kind of analysis?

11. What, specifically, are the differences between the DFT and the STFT? In what way is the STFT superior?

12. What capabilities do 1-D wavelets provide that were, up to now, not easily available with either conventional FDs or FTs?

13. Why are 1-D wavelets generating so much interest in so many disciplines? Do they offer the same promise in the biological sciences? Give some examples where you think wavelet analysis might be valuable.

14. What are the differences between the continuous wavelet transform (CWT) and the discrete wavelet transform (DWT)?

15. What is the difference between the DWT and its inverse IDWT?

REFERENCES CITED

Ahn, S. S., Lestrel, P. E. and Takahashi, O. (1999) The presence of sexual dimorphism in the human dental arches: A Fourier analytic study. *J. Dent. Res.* **78**:278.

Almgren, F. J. (1982) Minimal surface forms. *The Math. Intell.* **4**:164-172.

Almgren, F. J. and Taylor, J. (1976) The geometry of soap and soap bubbles. *Sci. Amer.* **235**:82-93.

Baughan, B. and Demirjian, A. (1978) Sexual dimorphism in the growth of the cranium. *Am. J. Phys. Anthrop.* **49**:383-390.

Blum, H. (1967) A transformation for extracting new descriptors of shape. **In** *Models for the perception of speech and visual form*. Wathen-Dunn, W. (Ed). Cambridge, Mass: MIT Press.

Blum, H. (1973) Biological shape and visual science. *J. Theoret. Biol.* **38**:205-287.

Blum, H. and Nagel, R. N. (1978) Shape description using weighted symmetric axis features. *Pattern Recog.* **10**:167-180.

Bookstein, F. L. (1977) The study of shape transformations after D'Arcy Thompson. *Math. Biosci.* **34**:177-219.

Bookstein, F. L. (1979) The line skeleton. *Comp. Graph. Imag. Proc.* **11**:123-137.

Bookstein, F. L. (1991) *Morphometric Tools for Landmark Data*. Cambridge: Cambridge University Press.

Borgefors, G., Nyström, I. and Di Baja, G. S. (1999) Computing skeletons in three dimensions. *Pattern Recog.* **32**:1225-1236.

Bracewell, R. N. (1989) The Fourier transform. *Sci. Amer.* **260**:86-95.

Brakke, K. A. (1992) The surface evolver. *Exp. Math.* **1**:141-165.

Brigham, E. O. (1974) *The Fast Fourier Transform*. New Jersey: Prentice-Hall.

Challis, R. E. and Kitney, R. I. (1991) Biomedical signal processing (in four parts) Part 2. The frequency transforms and their inter-relationships. *Med. Biol. Engr. Comput.* **29**:1-17.

Cooley, J. W. and Tukey, J. W. (1965) An algorithm for the machine calculation of complex Fourier series. *Math. Comput.* **19**:297-301.

Daegling, D. J. (1993) Shape variation in the mandibular symphysis of apes: An application of a median axis method. *Am. J. Phys. Anthrop.* **91**:505-516.

Davis, J. C. (1986) *Statistics and Data Analysis in Geology* (2nd Ed.). New York: John Wiley and Sons.

Diaz, G., Zuccarelli, A., Pelligra, I. and Ghiani, A. (1989) Elliptic Fourier analysis of cell and nuclear shapes. *Comp. Biomed. Res.*, **22**:405-414.

Diaz, G., Quacci, D. and Dell'Orbo, C. (1990) Recognition of cell surface modulation by elliptic Fourier analysis. *Comp. Meth. Prog. Biomed.* **31**:57-62.

Diaz, G., Cappai, C., Setzu, M. D., Sirugu, S. and Diana, A. (1997) Elliptical Fourier Descriptors of cell and nuclear shapes. **In** *Fourier Descriptors and their Applications in Biology*. Lestrel, P. E. (Ed.). Cambridge: Cambridge University Press.

Draper, N. R. and Smith, H. (1998) *Applied Regression Analysis* (3rd Ed.). New York: John Wiley.

Drolon, H., (1998) A fast algorithm to compute invariant wavelet descriptors. Tech. Report, LACOS, Le Havre University.

Drolon, H., Hoyez, B., Druaux, F. and Faure, A. (1999a) Wavelet analysis of sand grain roughness. *Comtes Rendus de l'Académie des Sciences, Sciences de la terre et des planètes*, **328**:457-461.

Drolon, H., Druaux, F and Faure, A. (1999b) Particles shape analysis and classification using the wavelet transform. *Math. Geol.* (submitted).

Ehrlich, R and Weinberg, (1970) An exact method for the characterization of grain shape. *J. Sediment. Petrol.* **40**:205-212.

Ehrlich, R. and Full, W. E. (1986) Comments on "Relationships among eigenshape analysis, Fourier analysis and analysis of coordinates" by F. James Rohlf. *Math. Geol.* **18**:855-857.

Ehrlich, R., Baxter Pharr, Jr. R. and Healy-Williams, N. (1983) Comments on the validity of Fourier descriptors in systematics: A reply to Bookstein *et al. Syst. Zool.* **32**:202-206.

Ferrario, V. F., Sforza, C., Gianni, A. B., Pogio, C. E. and Schmitz, J. (1990) Analysis of chewing movement using Elliptic Fourier descriptors. *Int. J. Adult Orthod. Orthognath. Surg.* **5**:53-57.

Ferrario, V. F., Sforza, C., Miani, A., Pogio, C. E. and Schmitz, J. (1992) Harmonic Analysis and clustering of facial profiles. *Int. J. Adult Orthod. Orthognath. Surg.* **7**:171-179.

Ferson, S., Rohlf, F. J. and Koehn, R. K. (1985) Measuring shape variation of two dimensional outlines. *Syst. Zool.* **43**:59-68.

Fujii, K. and Matsuura, Y., (1999) Analysis of the velocity curve for height by wavelet interpolation method in children classified by maturity rate. *Am. J. Hum. Biol.* **11**:13-30.

Full, W. E. and Ehrlich, R. (1982) Some approaches for location of centroids of quartz grain outlines to increase homology between Fourier amplitude spectra. *Math. Geol.* **14**:43-55.

Full, W. E. and Ehrlich, R. (1986) Fundamental problems associated with "eigenshape analysis" and similar "factor" analysis procedures. *Math. Geol.* **18**:451-463.

Halazonetis, D. J., Shapiro, E., Gheewalla, R. K. and Clark, R. E. (1991) Quantitative description of the shape of the mandible. *Am. J Orthod. Dentofac. Orthop.* **99**:49-56.

Hamming, R. W. (1973) *Numerical Methods for Scientists and Engineers* (2nd Ed.). New York: Dover Pub.

Harbaugh, J. W. and Merriam, D. F. (1968) *Computer Applications in Stratigraphic Analysis.* New York: John Wiley and Sons.

Hubbard, B. B. (1996) *The World According to Wavelets.* Wellseley, Massachusetts: A. K. Peters Ltd.

Inoue, M. (1990) Fourier analysis of the forehead shape of skull and sex determination by use of the computer. *Foren. Sci. Int.* **47**:101-112.

Johnson, D. R. (1985) Shape of vertebrae—An application of a generalized method. **In** *Normal and Abnormal Bone Growth: Basic and Clinical Research.* Dixon, A. D. and Sarnat, B. G. (Eds.). New York: Alan R. Liss.

Johnson, D. R. (1997) Fourier descriptors and shape differences: Studies on the upper vertebral column of the mouse. **In** *Fourier Descriptors and their Applications in Biology.* Lestrel, P. E. (Ed.). Cambridge: Cambridge University Press.

Kaesler, R. L. (1997) Phase angles, harmonic distance, and the analysis of form. **In** *Fourier Descriptors and their Applications in Biology.* Lestrel, P. E. (Ed.). Cambridge: Cambridge University Press.

Kennedy, D. N., Filipek, P. A. and Caviness, V. S. (1990) Fourier shape analysis of anatomic structures. **In** *Recent Advances in Fourier Analysis and its Applications.* Byrnes, J. S. and Byrnes, J. F. (Eds.). Dordrecht, Netherlands: Kluwer Academic Publishers.

Kieler, J., Skubis, K., Grzesik, W., Strojny, P., Wisniewski, J. and Dziedzic-Goclawska, A. (1989) Spreading of cells on various substrates evaluated by Fourier analysis of shape. *Histochem.* **92**:141-148.

Kincaid, D. T. and Schneider, R. B. (1983) Quantification of leaf shape with a microcomputer and Fourier transform. *Canadian. J. Bot.* **61**:2333-2342.

Kiryati, N. and Maydan, D. (1989) Calculating geometric properties from Fourier representation. *Pattern Recog.* **22**:469-475.

Kline, M. (1972) *Mathematical Thought from Ancient to Modern Times* (Vols. 1-3). Oxford: Oxford University Press.

Kuhl, F. P. and Giardina, C. R. (1982) Elliptic Fourier features of a closed contour. *Comp. Graph. Imag. Proc.* **18**: 236-258.

Lee, D. T. (1982) Medial axis transformation of a planar shape. *IEEE Trans. Pattern Anal. Mach. Intell.* PAMI-4:363-369.

Le Minor, J. M., Pister, L. and Kahn, E. (1989) Shape description on osteology using automatic Fourier analysis. *Arch. Anat. Hist. Embr. norm. et exp.* **72**:69-79.

Lestrel, P. E. (1974) Some problems in the assessment of morphological size and shape differences. *Yearbook Phys. Anthrop.* **18**:140-162.

Lestrel, P. E. (1980) A quantitative approach to skeletal morphology: Fourier analysis. *Soc. Photo. Inst. Engrs. (SPIE)* **166**:80-93.

Lestrel, P. E. (1982) A Fourier Analytic procedure to describe complex morphological shapes. In *Factors and Mechanisms Influencing Bone Growth*. Dixon, A. D. and Sarnat, B. G. (Eds.). New York: Alan R. Liss, Inc.

Lestrel, P. E. (1987) A new quantitative method for fitting growth data: Elliptical Fourier functions. *Am. J. Phys. Anthrop.***72**:224.

Lestrel, P. E. (1989a) Method for analyzing complex two-dimensional forms: Elliptical Fourier functions. *Am. J. Hum. Biol.* **1**:149-164.

Lestrel, P. E. (1989b) Some approaches toward the mathematical modeling of the craniofacial complex. *J. Craniofacial. Genet. Dev. Biol.* **9**:77-91.

Lestrel, P. E. (1997a), Morphometrics of craniofacial form: A Fourier analytic procedure to describe complex morphological shapes. In *Fundamentals of Craniofacial Growth*. Dixon, A. D., D. Hoyte, A. N. and Ronning, O. (Eds.). New York: CRC Press.

Lestrel, P. E. (1997b) *Fourier Descriptors and their Applications in Biology*. Cambridge: Cambridge University Press.

Lestrel, P. E. and Roche, A. F. (1976) Fourier analysis of the cranium in Trisomy 21. *Growth* **40**:385-398.

Lestrel, P. E. and Brown, H. D. (1976) Fourier analysis of adolescent growth of the cranial vault: A longitudinal study. *Hum. Biol.* **48**:517-528.

Lestrel, P. E., Kimbel, W. H., Prior, F. W. and Fleischmann, M. L. (1977) Size and shape of the hominoid distal femur: Fourier analysis. *Am. J. Phys. Anthrop.* **46**:281-290.

Lestrel, P. E. and Moore, R. N. (1978) *Macaca nemestrina:* A quantitative analysis of size and shape. *J. Dent. Res.* **57**:395-401. Erratum **57**:947.

Lestrel, P. E. and Roche, A. F. (1979) The cranial thickness in Down's syndrome: Fourier analysis. *Proc. 1st. Int. Cong. Aux. (Milan)* **1**:109-115.

Lestrel, P. E. and Sirianni, J. E. (1982) The cranial base in *Macaca nemestrina:* Shape changes during adolescence. *Hum. Biol.* **54**:7-21.

Lestrel, P. E. and Roche, A. F. (1986) Cranial base variation with age: A longitudinal study of shape using Fourier analysis: *Hum. Biol.* **58**:527-540.

Lestrel, P. E., Stevenson, R. G. and Swindler, D. R. (1988) A comparative study of the primate cranial base: elliptical Fourier functions. *Am. J. Phys. Anthrop.* **75**:239.

Lestrel, P. E., Engstrom, C., Chaconas, S. J. and Bodt, A. (1991) A longitudinal study of the human nasal bone in *Norma lateralis:* Size and shape considerations. **In** *Fundamentals of Bone Growth: Methodology and Applications.* Dixon, A. D., Sarnat, B. G. and Hoyle, D. A. N. (Eds.). Boca Raton Florida: CRC Press.

Lestrel, P. E. and Kerr, W. J. S. (1992) Shape changes due to functional appliances. *Calif. Dent. J.* **20**:30-36.

Lestrel, P. E. and Kerr, W. J. S. (1993) Quantification of functional regulator therapy using elliptical Fourier functions. *Europ. J. Orthod.* **15**:481-491.

Lestrel, P. E., Bodt, A. and Swindler, D. R. (1993) Longitudinal study of cranial base changes in *Macaca nemestrina. Am. J. Phys. Anthrop.* **91**:117-129.

Lestrel, P. E., Huggare, J. A., Ghiai, M., Matinfar, F. and Wolfe, C. A. (1994) Cranial base changes in shunt-treated hydrocephalics: Fourier descriptors. *J. Dent. Res.* **73**:444.

Lestrel, P. E. and Swindler, D. R. (1996) The numerical characterization of the primate cranial base: A comparative study using Fourier Descriptors. *Am. J. Phys. Anthrop.* Suppl. **22**:148.

Lestrel, P. E. and Huggare, J. A. (1997) Cranial base changes in shunt-treated hydrocephalics: Fourier descriptors. **In** *Fourier Descriptors and their Applications in Biology.* Lestrel, P. E. (Ed.). Cambridge: Cambridge University Press.

Lestrel, P. E., Read, D. W. and Wolfe, C. A. (1997) Size and shape of the rabbit orbit: 3-D Fourier descriptors. **In** *Fourier descriptors and their applications in biology.* Lestrel, P. E. (Ed.). Cambridge: Cambridge University Press.

Lestrel, P. E., Takahashi, O. and Ahn, S. S. (1999) An Analysis of dental arch form: A model based on Fourier descriptors. *J. Dent. Res.* **78**:278.

Lin, C. S. and Hwang, C. L. (1987) New forms of shape invariants from elliptic Fourier descriptors. *Pattern Recog.* **20**:535-545.

Lohmann, G. P. (1983) Eigenshape analysis of microfossils: A general morphometric procedure for describing changes in shape. *Math. Geol.* **15**:659-672.

Lohmann, G. P. and Schweitzer, P. N. (1990) On eigenshape analysis. **In** *Proceedings of the Michigan Morphometric Workshop.* Rohlf, F. J. and Bookstein, F. L. (Eds.). University of Michigan Museum of Zoology Special Pub. No 2.

Lu, K. H. (1965) Harmonic analysis of the human face. *Biometrics* **21**:491-505.

Mallat, S. (1989) A theory for multiresolution signal decomposition: The wavelet representation. *IEEE Trans. Pat. Anal. Mach. Int.* **11**:674-693.

Mok, E. C. and Boyer, A. L. (1986) Encoding patient contours using Fourier descriptors for computer treatment planning. *Med. Phys.* **13**:413-415.

Montanari U: Continuous skeletons from digitized images. *J. Assoc. Comp. Mach.*,1969, **16:** 534-549.

Moon, W., Lestrel, P. E., Engstrom, C. and Devincenzo, J. P. (1992) A quantitative approach for measuring Tx effects using functional appliance therapy: Fourier descriptors. *J. Dent. Res.* **71**:593.

Moon, W. (1992) *Mandibular changes with functional appliance therapy*. Unpublished MS thesis, Oral Biology, UCLA School of Dentistry.

Moon, W. (1997) A numerical and visual approach for measuring the effects of functional appliance therapy: Fourier descriptors. In *Fourier Descriptors and their Applications in Biology*. Lestrel, P. E. (Ed.). Cambridge: Cambridge University Press.

Murphy, G. F., Partin, A. W., Maygarden, S. J. and Mohler, J. L. (1990) Nuclear shape analysis for assessment of prognosis in renal cell carcinoma. *J. Urol.* **143**:1103-1107.

Nadler, M. and Smith, E. P. (1993) *Pattern Recognition Engineering*. New York: John Wiley and Sons.

Nafe, R., Kaloutsi, V., Choritz, H. and Georgii, A. (1992) Elliptic Fourier analysis of megakaryocyte nuclei in chronic myeloproliferative disorders. *Anal. Quant. Cytol. Histol.* **14**:391-397.

Newland, D. E. (1993) *An Introduction to Random Vibrations, Spectral and Wavelet Analysis* (3rd Ed.): U. K: Addison Wesley Longman Ltd.

Newland, D. E. (1994) Wavelet analysis of vibration, Part 2: Wavelet maps: J. *Vibration and Acoustics* **116**:417-425.

O'Higgins, P. and Williams, N. W. (1987) An investigation into the use of Fourier coefficients in characterizing cranial shape in primates. *J. Zool. Lond.* **211**:409-430.

Okamoto, V. E. (1994) *A numerical description of the rabbit eye orbit: Controls vs. experimental*. Unpublished MS thesis, Oral Biology, UCLA School of Dentistry.

O'Rourke, J. and Badler, N. (1979) Decomposition of three-dimensional objects into spheres. *IEEE Trans. Pattern Anal. Mach. Intell.* **PAMI-1**:295-305.

Osman, A. 1993 *A three dimensional analysis of the shape of the rabbit orbit: Fourier descriptors*. Unpublished MS thesis, Oral Biology, UCLA School of Dentistry.

Ostrowski, K., Dziedzic-Goclawska, A., Strojny, P., Grzesik, W., Kieler, J., Christensen, B. and Mareel, M. (1986) Fourier analysis of the cell shape of paired human urothelial cell lines of the same origin but of different grades of transformation. *Histochem.* **84**:323-328,

Oxnard, C. E. (1973) *Form and Pattern in Human Evolution*. Chicago: The University of Chicago Press.

Pavlidis T. (1978) A review of algorithms for shape analysis. *Comp. Graph. Imag. Proc.* **7**:243-258.

Parnell, J. N. and Lestrel, P. E. (1977) A computer program for fitting irregular two-dimensional forms. *Comp. Prog. Biomed.* **7**:145-161.

Persoon, E. and Fu, K. (1977) Shape discrimination using Fourier descriptors. *IEEE Trans. Pattern Anal. Mach. Intell.* **PAMI-8**:388-397.

Pesce Delfino, V. P., Colonna, M., Vacca, E., Potente, F. and Introna, J. (1986) Computer-aided skull/face superimposition. *Am. J. Foren. Med. Path.* **7**:201-212.

Pesce Delfino, V. P., Lettini, T. and Vacca, E. (1997) Heuristic adequacy of Fourier Descriptors: Methodologic aspects and applications in morphology. **In** *Fourier Descriptors and their Applications in Biology*. Lestrel, P. E. (Ed.). Cambridge: Cambridge University Press.

Pizer, S. M., Oliver, W. R. and Bloomberg, S. H. (1987) Hierarchical shape description via the multiresolution symmetric axis transform. *IEEE Trans. Pattern Anal. Mach. Intell.* **PAMI-9**:505-511.

Ramirez, R. W. (1985) *The FFT. Fundamentals and Concepts*. New Jersey: Prentice-Hall.

Ray, T. S. (1990) Application of eigenshape analysis to second order leaf shape ontogeny in *Syngonium podophyllum* (Araceae). **In** *Proceedings of the Michigan Morphometric Workshop*. Rohlf, F. J. and Bookstein, F. L. (Eds.). University of Michigan Museum of Zoology Special Pub No 2.

Ricco, R., De Benedictis, G., Giardina, C., Bufo, P., Resta, L. and Pesce Delfino, V. (1985) Morphometric analytical evaluators of lymphoid populations in noneoplastic lymph nodes. *Anal. Quant. Cytol. Histol.* **4**:288-293.

Rohlf, F. J. (1986) Relationships among eigenshape analysis, Fourier analysis and analysis of coordinates. *Math. Geol.* **18**:845-854.

Rohlf, F. J. (1990) Fitting curves to outlines. **In** *Proceedings of the Michigan Morphometric Workshop*. Rohlf, F. J. and Bookstein, F. L. (Eds.). University of Michigan Museum of Zoology Special Pub No 2.

Rohlf, F. J. and Archie, J. W. (1984) A comparison of Fourier methods for the description of wing shape in mosquitos *(Diptera: culicidae)*. *Syst. Zool.* **33**:302-317.

Sarnat, B. G. (1981) The orbit and eye: Experiments on volume in young and adult rabbits. *Acta Ophthalmol. Suppl.* **147**:1-43

Sarnat, B. G. and Shanedling, P. D. (1972) Orbital growth after evisceration or enucleation without and with implants. *Acta Anat. (Basel)* **82**:497-511.

Schwarcz, H. P. and Shane, K. C. (1969) Measurement of particle shape by Fourier analysis. *Sedimentology* **13**:213-231.

Schweitzer, P. N., Kaesler, R. L. and Lohmann, G. P. (1986) Ontogeny and heterochrony in the ostracode *Cavellina coreyell* from the Lower Permian rocks in Kansas. *Paleobiology* **12**:290-301.

Sneath, P. H. A. and Sokal, R. R. (1973) *Numerical Taxonomy*. San Francisco: W. H. Freeman and Co.

Straney, D. O. (1990) Median axis methods in morphometrics. **In** *Proceedings of the Michigan Morphometric Workshop*. Rohlf, F. J. and Bookstein, F. L. (Eds.). University of Michigan Museum of Zoology Special Pub No 2.

Strang, G., (1994) Wavelets. *Amer. Sci.* **82**:250-255.

Strojny, P., Traczyk, Z., Rozycka, M., Bem, W. and Sawicki, (1987) W. Fourier analysis of nuclear and cytoplasmic shape of blood lymphoid cells from healthy donors and chronic lymphocytic leukemia patients. *Anal. Quant. Cytol. Histol.* **9**:475-479.

Takahashi, O., Lestrel, P. E. and Ahn, S. S. (1999) A size and shape study of dental arch crowding versus non-crowding: Fourier analysis. *J. Dent. Res.* **78**:278.

Thomas, M. C., Wiltshire, R. J. and Williams, A. T. (1995) The use of Fourier descriptors in the classification of particle shape. *Sedimentology* **42**:635-645.

Tolstov, G. P. (1962) *Fourier series*. New Jersey: Prentice-Hall.

van Otterloo, P. J. (1991) *A Contour-oriented Approach to Shape Analysis*. New York: Prentice-Hall.

Wallace, T. P. and Mitchell, O. R. (1980) Analysis of three-dimensional movement using Fourier descriptors. *IEEE Trans. Pattern Anal. Mach. Intell.* **PAMI-2**:583-588.

Walker, J. S. (1999) *A Primer on Wavelets and their Scientific Applications*. Boca Raton, Florida: CRC Press.

Waters, J. A. (1977) Quantification of shape by the use of Fourier analysis: the Mississippian blastoid genus *Pentremites*. *Paleobiol.* **3**:288-299.

Webber, R. L. and Blum, H. (1979) Angular invariants in developing mandibles. *Science* **206**:689-691.

White, R. J. and Prentice, H. C. (1988) Comparison of shape description methods for biological outlines. In *Classification and Related Methods in Data Analysis*. Bock, H. H. (Ed.) North Holland: Elsevier Science Pub BV.

White, R. J., Prentice, H. C. and Verwijst, T. (1988) Automated image acquisition and morphometric description. *Can. J. Bot.* **66**:450-459.

Wolfe, C. A. (1997) Software for elliptical Fourier function analysis. In *Fourier Descriptors and their Applications in Biology*. Lestrel, P. E. (Ed.). Cambridge: Cambridge University Press.

Wunsch, P. and Laine, A. F. (1995) Wavelet descriptors for multiresolution recognition of handprinted characters. *Pattern Recog.* **28**:1237-1249.

Zahn, C. T. and Roskies, R. Z. (1972) Fourier descriptors for plane closed curves. *IEEE Trans. Comput.* **C-21**:269-281.

10. STRUCTURAL MORPHOMETRICS

Messen ist wissen.
To measure is to know.

Ernst Werner von Siemens (1816-1892)

10.1. INTRODUCTION

When one is interested in numerically characterizing the surface or internal structure of a form then one enters the realm of structural morphometrics. The numerical characterizations of the texture or surface patterning as well as the internal configuration of a form remain challenging endeavors and future advances will, undoubtedly, replace current approaches. The question of what is texture does not admit an easy answer. At the moment, formal definitions of texture are either non-existent or *ad hoc* in character (Nadler and Smith, 1993). Textures can range from completely regular to highly irregular. As Nadler and Smith indicate:

> Consideration of these two extremes leads to the formulation of two very different approaches to textural analysis. The study of basically regular patterns ... demands a technique that can uncover both the basic arrangement and the basic pattern used to generate it ... The analysis of basically irregular textures ... where there is significant random variation in both the components and their arrangement ... demands a completely different approach (Nadler and Smith, 1993:226).

That is, while the concept of texture seems intuitively apparent and visually readily recognizable, it has been quite difficult to define in quantitative terms. Thus, regardless of the above dichotomy, it is both this lack of definition as well as the extensive textural patterns that are present, that have held back significant development. Another problem that appears is that of scale. What is meaningful at one level becomes random and meaningless at another level. For example, consider a photograph. When greatly enlarged, only different-sized dots, either in color or in black and white, remain. So, at what observational level does an arrangement of recognizable objects become texture? We now need to leave such issues aside and focus on the morphometrics of texture.

Textural considerations are involved at least three levels. These are: [1] the surface texture or 'roughness' as seen in fabrics and metallurgy or the surface texture of biological forms which can range from regular to extremely varied; [2] the internal patterning underneath the surface displayed, for example, by the alignment of the bony spicules in the femur head or vertebrae; and [3] the physical and chemical properties, often displaying regularity such as the lattice structure of crystals. All these levels overlap to some extent and influence each other. Three approaches are currently available. Two of these are highly related, Fourier transforms and optical data analysis (or coherent optical processing). Another one is wavelets; specifically, 2-D wavelets. Textural analysis possesses special problems and tends to require analysis in 2-D hence the need for 2-D Fourier transforms and 2-D wavelets; and these are described in the next sections.

10.2. THE FOURIER TRANSFORM REVISITED

Earlier sections dealing with FDs, either the conventional FD (Section 9.3) or the elliptical formulation (Section 9.5), were based on Fourier's series. The Fourier transform, (FT), on the other hand, is a frequency-dependent list of amplitudes and phases, often shown in diagrammatic fashion (see Ramirez, 1985 for illustrations). It should be recalled that the FT (Section 9.6) is a mapping from the time or spatial domain into the frequency domain, exactly analogous to FDs. In fact, the Fourier series is simply a special case of the FT (Brigham, 1974; Challis and Kitney, 1991). The FT frequency domain contains exactly the same information that is present in the original observed function. It is now described with complex numbers which represent computational advantages as well as leading to the fast Fourier transform (FFT) briefly described in Section 9.6.2.

A common procedure in textural analysis is to use the 2-D FT to generate a power spectrum. It will be recalled that the power spectrum is simply computed as the square of the amplitudes and then summed over the Fourier series (Section 9.3.3). However, it is important to realize that the power spectrum does not contain information on phase. Nevertheless, phase is an important property with textural analysis, as textures with identical power spectra can appear quite distinct from each other if they differ in their phase spectra (see Figure 5.19 in Nadler and Smith, 1993). Interestingly, form (*i.e.,* outline) in contrast to texture, seems to be especially affected by the phase spectrum in contrast to power spectrum.

As already mentioned, The FFT is particularly useful in cutting down the excessive CPU times that are experienced as the number of harmonics increase. In the 2-D FT case, this CPU time can become significantly lower (Section 9.6.2). That is, instead of roughly $N^2 M^2$ computations, the FTT now take about $NM \cdot \log_2(NM)$ calculations, where the image consists of N rows and M columns with both N and M set up in terms of the required powers of two (Nadler and Smith, 1993).

10.3. OPTICAL DATA ANALYSIS

A FT related approach is optical data analysis or as it is sometimes termed, coherent optical processing (Oxnard, 1973; Lugt, 1974). Coherent refers to the use of a laser light source. Instead of numerically computing either the Fourier series or the transform, the procedure is now carried out using optical means. This makes the power spectrum generated from the optical data analysis, an analog computation in contrast to a digital one.

10.3.1. The One-Dimensional Optical Power Spectrum

Consider the passage of a light beam through a prism. The prism acts as a frequency analyzer and transforms or separates the waveform (white light) into components at different wavelengths (the color spectrum). This is equivalent to the transformation from the spatial domain (white light) to the frequency domain (color spectrum). In an equivalent fashion, the use of an optical data analyzer utilizes the power spectrum derived from an image and breaks it down into constituent parts to be analyzed, in effect, a harmonic analyzer. If one scans along, for example, a horizontal line of a black and white image, one can generate a complex 1-D waveform, which can be then transformed from the spatial domain into the frequency domain in the usual fashion. In this case, the peaks represent bright regions in the scan and the valleys, dark regions. This 1-D scan can then be computationally expressed as a power spectrum:

$$f(x') = \left| \int f(x) \exp^{-j\omega x} dx \right|^2, \qquad [10\text{-}1]$$

where $f(x)$ is the 1-D input signal as a transverse wave, ω is a cosine vector, j is a complex constant and, $f(x')$, the power spectrum, is the square of the Fourier transform (Oxnard, 1973). Note that equation [10-1] is squared because it represents power instead of magnitude (amplitude). See Section 9.3.3. Rather than computing equation [10-1], the process is performed optically in an analogous way to the prism example above. However, the data usually analyzed with the optical approach consists of 2-D images or 'pictures' such as radiographs, which contain black particles and white spaces or pores that extend in both the x- and y-directions, so that the 1-D approach is insufficient; that is, this requires a 2-D approach.

10.3.2. The Two-Dimensional Optical Power Spectrum

Two-dimensional images can be described with x, y coordinates (pixels) and the additional information of gray scale values available for each pixel, together they can then be viewed as a 2-D optical FT. Utilizing a laser source and a series of lenses one can produce the transform and record the amplitudes as an image on film (Oxnard, 1973).

The power spectrum of the optical FT represents the transformation from the spatial domain to the frequency domain. This in optical terms, as stated earlier, can be visualized on film as a white 'cloud' or pattern of radiating lines or white dots arranged in a polar coordinate fashion around a central axis.

$$f(x', y') = \left| \int \int f(x, y) \exp^{-j[\omega x + \omega y]} dx \cdot dy \right|^2, \qquad [10\text{-}2]$$

where $f(x, y)$ refers to the 2-D input signal of an image or picture, ω is a cosine vector in the x and y directions, j is a complex constant and $f(x', y')$ is the square or power of the 2-D Fourier transform (Oxnard, 1973). For discrete digital images, the 2-D DFT is utilized and the integrations are replaced by the usual summations. These summations are then truncated at the edges of the image of interest. However, in the analog case, here, image analysis is accomplished using optical means.

10.3.3. Two-Dimensional Optical FT Applications

An early study using the optical Fourier approach, explored the numerical characterization of particles such as metal powders (Kaye and Naylor, 1972). They initially presented the diffraction pattern (power spectrum) displayed by regular figures such as a square, ellipse and circle, as well as some irregular ones. They concluded that the distinctiveness of those images should make the optical method feasible for the characterization of the shape of particle images. Power and Pincus (1974) applied the optical FT to petrographic sections and demonstrated how the diffraction patterns produced by the transform correlated with the actual thin section photomicrographs.

In a chemical context, the numerical description of lattice images derived from electron microscopy and tunneling microscopy could, potentially be efficiently exploited using optical FTs (Ourmazd, *et al.*, 1989; see also Steward, 1988).

Oxnard, (1973; 1997) applied optical data analysis in numerous contexts dealing with the internal structure of bone. In an attempt to clarify the relationship between the cancellous architecture of bone and the stresses placed on it, the optical FT was used to detect differences in the cancellous network of vertebrae. The method is applicable to transparencies of sections, CT[93] scans and radiographs. Utilizing a laser source to produce the optical FTs, Oxnard and his colleagues, demonstrated specific differences in patterning that could be associated with the actual internal structure of the vertebral bodies in question (Buck, 1990). They found that the FT produced information relating to the size, direction and spacing of the cancellous bony elements.

When the optical FT was used to look at the cancellous bone patterns in human and non-human primate vertebrae, significant differences appeared. These differences were not readily apparent on visual examination of the vertebrae. In brief, the human vertebral body displayed a generalized orthogonality or 'cross' form (with horizontal and vertical elements predominant). This pattern was also present in gorillas, chimpanzees and bonobos (pygmy chimpanzees). However, it was absent in the orangutan. The pattern in the orangutan was found to be circular rather than cross-like. Nevertheless, it is to be emphasized that these optical FT differences are qualitative.

These results suggested that bipedality in humans, and incipient bipedality in those apes mentioned, strongly influenced the architectural pattern of the human vertebrae in contrast to the orangutan, which due to its arboreal habitat has a peculiar quadrupedal pattern of locomotion (Oxnard, 1980; 1982; 1997). As a consequence, necessarily different mechanical stress and strain loads are imparted to the human and African apes, in contrast to the orangutan. Unfortunately, no data was available on another ape, the gibbon.

In another optical FT study, limited to human materials, the cancellous structure of the second lumbar vertebrae could be clearly distinguished from the fourth, while visually these internal differences were indistinguishable. However, these differences did appear in scanning electron micrographs, from which functional implications could be inferred (see Oxnard, 1997). Differences were also obtained between males and females at different ages as well as distinct patterns at different regions within the same vertebrae (Oxnard and Buck, 1990).

---

[93] Computerized tomography.

Interestingly, vertebral body differences were also found between normal and osteoporotic individuals. This study was subsequently expanded. It became apparent that, in general, the cross-like pattern characteristic of a non-osteoporotic condition, gives way to a less orthogonal, more football-like structure with advancing age. Figure 10.1 (adapted from Oxnard, 1997) shows the optical Fourier transforms of the upper right quadrants of lumbar vertebrae, with: [A] displaying a young individual and [B] an old individual. This pattern was present in both males and females. However, in females, it occurred earlier as well as being more pronounced (Oxnard, 1997). Using principal components, a somewhat surprising result was that a comparison of young males and young females showed a complete separation with no overlap. Reasons for this state of affairs were not entirely clear (Oxnard, 1997).

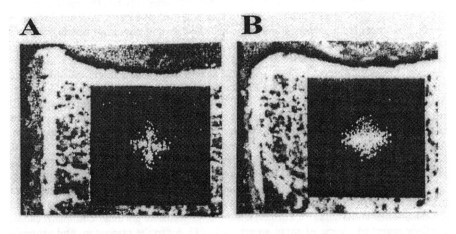

Figure 10.1. Optical Fourier transforms of the lumbar vertebrae.

Finally, of particular clinical interest was the fact that a few young females, who did not exhibit any of the overt clinical signs of osteoporosis, were nevertheless symptomatic for the disease using the optical FT. That is, they showed a significant loss of the cross-like form and instead displayed the football-like form associated with older ages (Fig. 10.1, on the right). This suggests the possibility of using the optical FT as a predictor for routine screening of osteoporosis.

Nevertheless, one of the drawbacks of the purely optical approach in terms of the power spectra has to do with interpretation. That is, how to give a quantitative analysis to what is generally a qualitative image composed of a cloud of white points on a black field, although some approaches have been put forward (Nadler and Smith, 1993). While Oxnard and colleagues were able to assign functional interpretation to their images, this is not always possible. Hence, while an explanation of differences seen in textural patterns is often possible in qualitative terms using the optical FT, few quantitative procedures are available that can be used to directly convert the pattern into numerical data[94]. Finally, with respect to purely textural data such as fabrics or thin sections of rocks, etc., it has been said that the presence of:

> ... a mosaic of textures or differentially oriented anisotropic textures ... the power spectrum will not be of much use (Nadler and Smith, 1993:258).

While the issues addressed by pattern recognition theorists may seem to necessarily differ from those of central concern to biologists, they, nevertheless, share many common attributes. It is hoped that some of these problems will be alleviated with the increasing utilization of 2-D wavelets, which are taken up in the next section.

10.4. TWO-DIMENSIONAL WAVELETS

Wavelets were briefly introduced in Chapter 9, albeit in a mostly non-technical fashion. A detailed understanding of the theory underlying wavelets requires considerable mathematical sophistication beyond the scope or purpose of this volume. Besides Fourier analysis, some knowledge of the theory of vector spaces, Riemann integrals, Lebesgue theory, Hilbert spaces, Banach spaces, etc., is generally required. Those interested in acquiring this background are strongly urged to read the works by Mallat (1989), Chui (1992), Daubechies (1992), Meyer (1992), Koornwinder, (1993) and Walker (1999) to name only a few; although, very little of this material can be considered particularly elementary in nature. Walker (1999) is, perhaps, one exception. The first four references mentioned have become classics in the field. Finally, mention should be made of a very readable book that was intended to introduce wavelets to the scientifically interested non-specialist (Hubbard, 1996; 1998).

One can readily carry out wavelet analyses using one of the numerous computer programs that have recently appeared. Wavelet toolboxes are now available for Mathematica™ and Matlab™. Further, a number of websites are available which can be fruitfully searched. Some of these sources are: [1] A free electronic journal, *Wavelet Digest*, which can be obtained from **http://www.wavelet.org**, [2] another source for information is **http://www.amara.com/current/wavelet.html**, [3] a free introductory wavelet package called FAWAV, available at **http://wwwcrcpress.com/edp/download**, [4] another free program can be found at **http://home.zcu.cz/~filip/wstudio.html**, and [5] one of the wavelet toolboxes that are available for Matlab™, this one is available from **http//www-dsp.rice.edu/software/RWT**. Numerous other sources are available on the Internet with some diligent searching.

10.4.1. Wavelet Analysis of Two-Dimensional Images

The use of the 2-D discrete wavelet transform (DWT) to extract frequency information from textures is similar to the FTs in principle, except, again, with the advantage that positional information is now being maintained (Section 9.7). That is, one can identify the localized information of the frequency components in the time/boundary representation, a property not possible with conventional FTs. The continuous version of the 2-D wavelet transform, the CWT, can be set up as:

---

[94] But see Oxnard (1997) for a number of quantitative approaches for numerically characterizing the image generated by the optical FT

$$C(j_x, k_x, j_y, k_y) = \int_{-\infty}^{\infty} \int_{-\infty}^{\infty} f(u,v)\Psi(j_x, k_x, j_y, k_y, u, v) du\ dv, \qquad [10\text{-}3]$$

where j_x and j_y refer to the scaling factors, or levels; k_x and k_y to the position factor; $f(u, v)$ is the 2-D function to be decomposed into wavelets using multiresolution analysis; and the term $\Psi(j_x, k_x, j_y, k_y, u, v)$ again symbolically refers to the basis function of the wavelet. Finally, u and v refer to the x- and y-axes respectively. Therefore, from equation [10-3], the wavelet amplitude coefficients, $C(j_x, k_x, j_y, k_y)$, are a now a function of *both* scale and position but separately along the x and y-axes.

The 2-D discrete case, the DWT, can best be viewed with an example. Consider a photograph with one central object, a horse. Figure 10.2, adapted from Hubbard (1996), shows this photographic image decomposed into three slices: [A] low frequencies, [B] medium frequencies and [C] high frequencies; although, there is image overlap in terms of these frequencies. The low frequency information derived from the 2-D wavelet tends to be focused on broad, global aspects of the image. Here the horse outline can be discerned but it is quite fuzzy. The middle frequencies begin to display more of the image detail but the image is still somewhat fuzzy. The high frequencies, in contrast, provide a clear picture of the details. In this case the texture of the coat and the hair of the horse's mane. This is precisely analogous to FDs where it will be recalled that the first few harmonics account for the global aspects of the form, while the higher harmonics are responsible for the fine details in the boundary outline (Section 9.3.4).

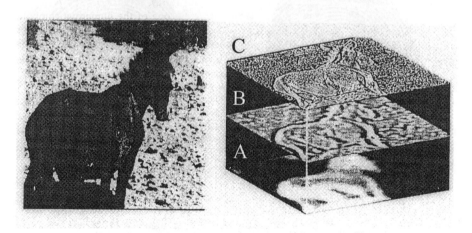

Figure 10.2. Image of a horse and its associated 2-D wavelet decomposition.

Another way to view Figure 10.2 is in terms of filters. That is, the original signal (horse image) has been passed through a triplet of low-pass, mid-pass and high-pass filters to get three new signals. This approach is used to build a set of discrete wavelet coefficients that are a function of scale and position of the fundamental analyzing wavelet (refer to Section 9.7.1 for the 1-D case). Thus, the low-pass filter is equivalent to level 1 (scale=2^1) in the multiresolution analysis and the high-pass filter can be considered at the highest level (scale=2^n) achievable, which is dependent on the sampling of the image. The results of this successive filtering process, a multiresolution analysis, allows the construction of a wavelet decomposition tree commonly seen in the literature and generated by most wavelet routines, for example, Matlab™.

Image analysis using the 2-D DWT requires a number of steps: [1] a decomposition level (multiresolution analysis) is selected, [2] a specific wavelet is chosen for analysis, Haar, Daubechies, etc., [3] the image is decomposed into wavelet coefficients and [4] using the IDWT the image can then be re-created, at will, at various levels.

10.4.2. Two-Dimensional Wavelet Applications

Application of 1-D wavelets, including signal processing, the de-noising of data, EEG diagnostics, speech synthesis etc., were briefly mentioned in Section 9.7. Two-dimensional wavelets, on the other hand, find uses in the characterization of images, and particularly the compression of images for future retrieval.

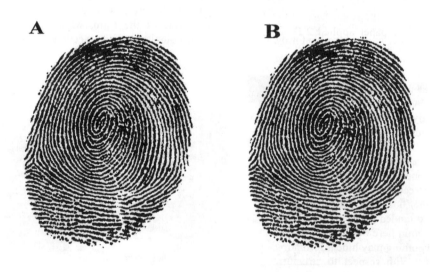

Figure 10.3. Compression of fingerprint data using wavelets.

An example of compression is the now familiar fingerprint data for the U. S. Federal Bureau of Investigation (FBI). Figure 10.3 is adapted from Brislawn (1996). The fingerprint on the left, [A], is the original image measuring 768 x 768 pixels (=589,824 bytes) as a JPEG file. In contrast, the compressed fingerprint, [B], is a reconstructed image that is only 37,702 bytes, a compression ratio of 18 to 1. This compression is with virtually no loss of the detail found in the original image. The standard used to create this compression is a DWT algorithm referred to by Brislawn (1996) as a Wavelet/Scalar Quantization (WSQ). The procedure starts with a WSQ *encoder*, which uses the DWT to generate 64 frequency levels, in effect, a multiresolution analysis. The DWT output is then truncated (or 'quantized') by the scalar quantization step. Finally, the quantized DWT output is Huffman-coded to minimize the number of bits to be transmitted. To re-generate the image after compression, the WSQ *decoder* is utilized, which undoes the Huffman coding, maps the DWT coefficients back to original data values and invokes the inverse DWT (Brislawn, 1996).

Another application, in a similar vein to fingerprints, is the recent work on characterizing iris patterns to verify a person's identity (Wilson, 1999). Ophthalmologists have established that the human iris remains unchanged for decades. A video image was taken of the iris and its inner and outer borders located. An algorithm, based on a line integral, was developed which searches the image looking for circular contours. Once located, the inner and outer borders, as well as the eyelash margins are fitted with splines. The next step is to construct a multiresolution analysis using 2-D Gabor wavelets to extract information from the iris. A 256-byte image is then developed that is a unique description of a person's iris. Utilizing a RISC-based CPU, identification can be obtained in 100 ms. Additionally, this time can be reduced still further since an accurate identification can still be made even with a 33% mismatch of the image with its database.

One paper applied 2-D wavelets to images in a biological context (Kiltie, *et al*, 1995). Their purpose was to provide a wavelet metric that could be applied to textural discrimination incorporating recent advances in computational vision and psychobiology. Their method was based on the wavelets' ability to allow localized analysis of orientation and frequency of visual textures. Examples of textural discrimination included the fur coat of tigers and zebras. Another recent paper applied multi-scaling analysis to characterize the pattern of shell growth due to the accumulation of $CaCO_3$ (Toubin, 1999).

If wavelets, either in 1-D or in 2-D, are to become a major tool for the analysis of form in a biological context, then the boundary outline localizations found must be mapped onto the homology space. For those situations where boundary discontinuities are readily apparent, this may not prove to be a problem. However, the presence of minor boundary perturbations, possibly meaningful in a biological context, may become confounded with error or noise. De-noising or smoothing, presumably, being one solution here but pitfalls remain. Over smoothing may loose critical information, under smoothing may hide information.

With respect to structural or textural considerations, other issues emerge. As demonstrated, 2-D wavelets allow excellent compression of fingerprint data (18:1) making storage requirements of the over 200+ million cards containing the fingerprint data, more manageable. Reducing storage requirements is a clearly important practical consideration, of that there is little question. In addition, the ability to uniquely identify the human iris is noteworthy and useful in a practical sense. Both of these accomplishments, however, were driven by largely practical considerations and technology/research supplied answers. From a biological context there remains a more fundamental issue at stake; namely, the old and enduring relationship of form with function.

What remains to be addressed is a return to one of the central tenets of this volume: the relationship of form with process. That is, *relating the form to the underlying process that is responsible for it*. One problem, for instance, with the fingerprint data, would be how to numerically characterize the visual information composed of the complex textural minutiae of the ridges in a biologically meaningful manner. Similarly, for the iris data, the question that arises is whether the patterns seen can be related to biological processes. In other words, morphometric characterization is only the initial step; we need to subsequently focus on identifying those process(s) involved, which determine fingerprint, iris data, etc. Clearly, these are continuing areas of challenge as well as opportunity for the next century of biologists.

Morphometrics from a general biological standpoint, as well as structural morphometrics specifically, is in transition. There seems to be a shifting from the initial emphasis primarily concerned with numerical description, and a turn toward the development of increasingly sophisticated models of form. Models, which attempt to closely mirror biological variability on one hand, and a greater awareness of the importance of morphometrics in a biological context on the other hand. This should bring us closer to the goal of explaining process. In the words of D'Arcy Thompson:

> We begin by describing the shape of an object in the simple words of common speech: we end by defining it in the precise language of mathematics. ... Lastly, and this is the greatest gain of all ... we pass from the concept of Form to an understanding of the Forces which gave rise to it (Thompson, 1915:857).

KEY POINTS OF THE CHAPTER

This concluding chapter has focused on image analysis, which was characteristically composed of 2-D datasets. Two approaches were discussed. The first one was optical data analysis and the second one was 2-D wavelets. Both methods were set within the framework of textural analysis and comprised some of the methods of structural morphometrics.

CHECK YOUR UNDERSTANDING

1. List some of the characteristics inherent in textural images. Why do they remain a challenge in terms of numerical characterization?

2. How does observation level relate to textural images? What are some of the consequences of this for the biological sciences in general?

3. What is meant by feature extraction as used in pattern recognition analysis and how does this concept apply in a biological context.

4. Compute the time required for both the 1-D and 2-D FFT. Display the differences in a graph. Hint: See Sections 9.6.2 and 10.2.

5. Given the optical FT data applications discussed in the chapter, can you think of other areas where the optical FT might prove useful?

6. Can you think of some ways in which the visual information derived from the optical FT can be described in numeric rather than qualitative terms? Can you consider of way to improve on these? Hint: See Oxnard, 1997.

7. What advantages does the 2-D DWT have over the 2-D optical FT?

8. What is a wavelet decomposition tree and how is it created?

9. What is the difference between a multiresolution analysis in the 1-D and the 2-D wavelet transform? How would you show these differences graphically?

10. After carefully considering question #5, can you think of some numeric methods that would also apply to the output of 2-D wavelet transforms?

11. Define the following pair of terms: *Decomposition/Reconstruction, Analysis/ Synthesis, encoder/decoder, DWT/IDWT.* What do they share in common?

REFERENCES CITED

Brigham, E. O. (1974) *The Fast Fourier Transform.* New Jersey: Prentice-Hall.

Brislawn, C. (1996) The FBI fingerprint image compression standard. URL: **http://www.c3.lanl.gov/~brislawn/FBI/FBI.html**.

Buck, A. M. (1990) Variation within trabecular networks of the lumbar vertebrae in modern humans. *Am. J. Phys. Anthrop. Suppl* **14**:55.

Challis, R. E. and Kitney, R. I. (1991) Biomedical signal processing (in four parts) Part 2. The frequency transforms and their inter-relationships. *Med. Biol. Engr. Comput.* **29**:1-17.

Chui, C. K. (1992) *An Introduction to Wavelets.* New York: Academic Press.

Daubechies, I. (1992) *Ten Lectures on Wavelets.* Philadelphia, PA: SIAM.

Hubbard, B. B. (1996) *The World According to Wavelets.* Wellseley, Massachusetts: A. K. Peters Ltd.

Hubbard, B. B. (1998) *The World According to Wavelets.* (2nd Ed.) Wellseley, Massachusetts: A. K. Peters Ltd.

Kaye, B. H. and Naylor, A. G. (1972) An optical information procedure for characterizing the shape of fine particle images. *Pattern Recog.* **4**:195-199.

Kiltie, R. A., Fan, J. and Laine, A. F. (1995) A wavelet-based metric for visual texture discrimination with applications in evolutionary ecology. *Math. Biosci.* **126**:21-39.

Koornwinder, T. H. (1993) *Wavelets: An Elementary Treatment of Theory and Applications.* Singapore: World Scientific.

Lugt, A. V. (1974) Coherent optical processing. *Proc IEEE* **62**:1300-1318.

Mallat, S. (1989) A theory for multiresolution signal decomposition: The wavelet representation. *IEEE Trans. Pattern Anal. Mach. Int.* **11**:674-693.

Meyer, Y. (1992) *Wavelets and Operators*. Cambridge: Cambridge University Press.

Nadler, M. and Smith, E. P. (1993) *Pattern Recognition Engineering*. New York: John Wiley and Sons.

Ourmazd, A., Taylor, D. W., Bode, M. and Kim, Y. (1989) Quantifying the information content of lattice images. *Science* **246:**1571-1577.

Oxnard, C. E. (1973) *Form and Pattern in Human Evolution*. Chicago: The University of Chicago Press.

Oxnard, C. E. (1980) The analysis of form: without measurement and without computers. *Am. Zool.* **20:**695-705.

Oxnard, C. E. (1982) The association between cancellous architecture and loading in bone: An optical data analytical view. *The Physiol.* **25:**Suppl:S37-S40.

Oxnard, C. E. and Buck, A. M. (1990) Age and sex differences in cancellous patterns in human vertebral bodies: a matter of osteoporosis? *Am. J. Phys. Anthrop., Suppl.* **14:**130.

Oxnard, C. E. (1997) From optical to computational Fourier transforms: The natural history of an investigation of the cancellous structure of bone. **In** *Fourier Descriptors and their Applications in Biology*. Lestrel, P. E. (Ed.). Cambridge: Cambridge University Press.

Power, P. C. and Pincus, H. J. (1974) Optical diffraction analysis of petrographic thin sections. *Science* **186:**234-239.

Ramirez, R. W. (1985) *The FFT. Fundamentals and Concepts*. New Jersey: Prentice-Hall.

Steward, M. (1988) Computer image processing of electron micrographs of biological structures with helical symmetry. *J. Elec. Micro. Tech.* **9:**325-358.

Toubin, M., Dumont, C., Verrecchia, E. P., Laligant, O., Diou, A., Truchetet, F. and Abidi, M. A. (1999) Multi-scale analysis of shell growth increments using wavelet transform. *Comp. Geosci.* **25:**877-885.

Walker, J. S. (1999) A Primer on Wavelets and their Scientific Applications. Boca Raton, Florida: CRC Press.

Wilson, A. (1999) Wavelet analysis ensures accurate recognition of iris patterns. *Vision System Design* **4:**30-35

Thompson, D. W. (1915) Morphology and mathematics. *Trans. Roy. Soc. Edinburgh* **50:**857-895.

EPILOGUE

This volume has attempted to deal with two of the more basic challenges of modern biology. Namely, [1] how to numerically characterize the complex visual information that resides in all morphological forms and [2] how to relate form to process. These remain significant and challenging tasks, particularly the former, since how this visual information is assessed may have a direct bearing on explanations of biological process.

Viewed broadly, three distinct aspects of form can be recognized. These are: [1] the location of 'homologous points' on and within the boundary; [2] the shape or outline of the boundary; and [3] the texture lying on the surface or internal to the surface. Because of personal research predilections with respect to the numerical characterization of the boundary outline, the emphasis on occasion here has been on structures in the human craniofacial complex. However, the issues raised are applicable to all biological forms.

Of the various morphometric procedures discussed, it has become increasingly apparent that many contain serious constraints that limit their applicability, except, perhaps, in very simple cases. While a simple solution may be perfectly appropriate and desirable in a particular research instance, the information being elicited is also going to be necessarily equally limited. The use of the conventional metrical approach (CMA) is a case in point, since by focusing on a few subjectively chosen and isolated aspects of the form, there is always the potential that critical elements will be missed. This can have unintentional consequences in terms of interpretation and analysis of results. Although, CMA still represents the most widely used numerical procedure, it provides an incomplete mapping of the form domain to the measurement domain, with the result that much of the visual information of potential biological significance is lost. The recognition of this incomplete mapping has led to some of the alternate procedures discussed in this volume.

While these alternate procedures can be considered improvements in various ways over CMA, none of them can be considered complete models of form, in and of themselves. For example, if one is centrally concerned with shape or textural analysis one is inescapably faced with the limitations of these methods. Thus, all of the recent approaches still fall short of satisfying the need for a complete morphometric model of measurement, in the sense that all three aspects mentioned above are incorporated. Toward that end, a more sophisticated, if heuristic, formal model was proposed in Chapter 6.

In contrast, a working model that attempts to addresses at least one part of this problem; namely, the mapping of the 'homologous point' information into the boundary outline using EFFs, was proposed in Chapter 9, Section 9.6. At this moment however, there seems to be a total absence of data that relates the internal texture lying within the boundary to the boundary outline.

Of all of the procedures so far available, wavelets currently offer the most promise for characterizing both boundary and textural aspects of form; although, at the moment

247

there is a virtual absence of such biologically-based papers. Moreover, even here problems remain evident. To begin with, there are no models yet available that treat size, shape and textural considerations jointly. While wavelets are able to begin the task of capturing an appreciable percentage of the information contained within the visual 2-D image, how to best access it, remains problematic. That is, one can be overwhelmed by the sheer magnitude of the numerical information available in an image, requiring efficient reduction (smoothing, filtering etc.) methods that do not loose critical data. Moreover, most methods do not even begin to deal adequately with 3-D morphologies. Assessments reasonably close to a 100% of the visual information present in morphological forms remain a challenge for the 21st century.

We end by returning to the question raised in the introduction; namely, the relationship between morphology and process, and the role that morphometrics plays in it. Historically once the emphasis in scientific endeavors shifted to the elucidation of process, the observation of phenomena and the recording of facts turned increasingly quantitative, a departure from the largely descriptive approach used previously. Nevertheless, the numerical quantification of form remains in its infancy in the biological sciences. As Read has noted:

> Directly or indirectly, form is central to our understanding of biological and genetic processes. The form mediates between internal genetic information and external environment; it is the means by which genetic information is evaluated and acted upon by natural selection (Read, 1990:417).

Considering just one example of process from a biological standpoint, it has been implicitly obvious for a long time now, that the growing biological form not only changes its size but also its shape in complex, and at times, unpredictable ways. Moreover, the biological form is subjected to countless forces during its life cycle, some known and some potentially still unknown. Some of these forces are known to act concomitantly. These forces include long-term, time-dependent processes such as evolution, as well as shorter-term ones occurring during growth and development. In addition, functional forces such as biomechanical loadings imposed on the individual during the life cycle and environmental stresses, cannot be ignored. Thus, quantification of the form, at all scales, must remain as one of the essential first steps in the elucidation of process in the biological sciences:

> The issue of how to precisely and completely describe the information that resides in all morphological forms, remains a major challenge. Moreover, how the visual information present in the biological form is characterized by a mathematical representation, has a direct bearing on subsequent explanations of biological process. (Lestrel, 1997).

REFERENCES CITED

Lestrel, P. E. (1997) *Fourier Descriptors and their Applications in Biology*. Cambridge: Cambridge University Press.

Read, D. W. (1990) From multivariate to qualitative measurement: representation of shape. *J. Hum. Evol.* **5**:417-429

APPENDIX I. EFF23: A COMPUTER PROGRAM

A.1.1. INTRODUCTION

The following computer routines are intended to facilitate the numerical characterization of complex and irregular 2-D and 3-D boundary outlines commonly encountered in the biological sciences. EFF23, Version 4.0 (November, 1999) makes use of the elliptical Fourier function (EFF) algorithm developed by Kuhl and Giardina (1982). The 3-D computation viewed as a curve in space (not a solid/volume), is an extension of the Kuhl and Giardina algorithm. The programmatic extensions to 3-D, etc. were devised and implemented by C. Wolfe, D. Read and P. Lestrel during 1992—1995. It has taken several years to work out the algorithms for 2-D and 3-D orientation. The methodology and programming dealing with crossings in an outline (rarely handled by other software) and the formulation and inclusion of homologous points into the boundary outline were developed by C. Wolfe with assistance from P. Lestrel and D. Read and was completed by 1997. Version 4.0, which facilitates batch processing, an automation feature that greatly speeds up the computation tasks, was completed by C. Wolfe over 1998-1999. Text figures 7.1, 7.2, 9.4 and 9.5 in this volume were created from EFF generated data.

A.1.2. BRIEF OVERVIEW OF EFF23

The multiple functions of EFF23 are shown in Appendix II, Figure A.1. The researcher starts with a sample and generates a point system protocol that is uniform across specimens. Three essential steps are required: [1] each specimen is digitized to create the initial digitized or observed form containing (x, y) or (x, y, z) data; [2] the digitized form is then used to compute EFF coefficients for each harmonic and [3] the digitized form is also used to separately calculate predicted points (either homologous or evenly spaced). These three steps are the basis for all subsequent computations. Provision is also available at this time for orienting the form with the coordinate system and for standardizing for size. Once the EFF coefficients have been computed for the sample of digitized forms, one can then compute amplitudes, phase angles, and power spectra. These are displayed in Appendix II, Figure A.2. In addition, from the computed coefficients, a number of ellipse parameters can be obtained (Appendix II, Fig. A.3). From the computed predicted points, various distance measures can be generated such as centroid to outline distances, etc., (Appendix II, Fig. A.3). In addition, statistical means, standard deviation and standard error of the means, for all computed parameters are also available. Plotting of the forms, observed as well as predicted, is available. Finally, an export module allows the exporting of data to other programs such as spreadsheets, statistical programs and graphic packages.

APPENDIX II. EFF23 PROGRAM FLOWCHARTS

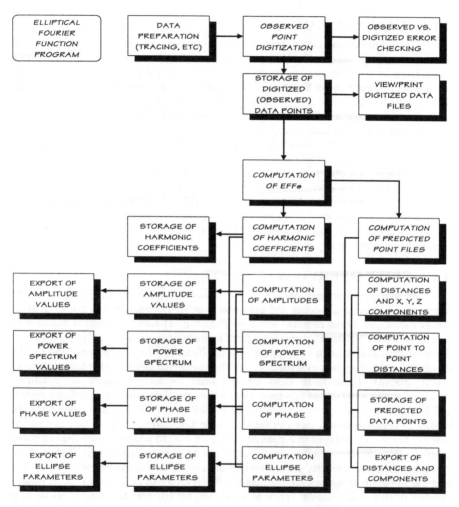

Figure A.1. Flowchart of data derived from EFF analysis.

Note: Boxes with text in italics refer to three essential steps from which all other computations are derived. These steps: [1] digitized data, [2] harmonic coefficients and [3] predicted point files. These three steps must be saved and archived for future use. All other computations can be generated at any time from these three datasets.

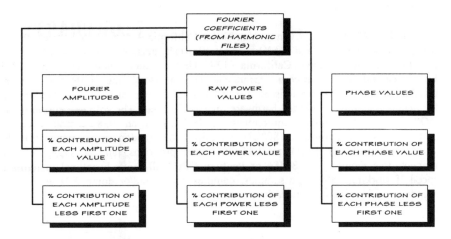

Figure A.2. Computation of amplitude, power and phase.

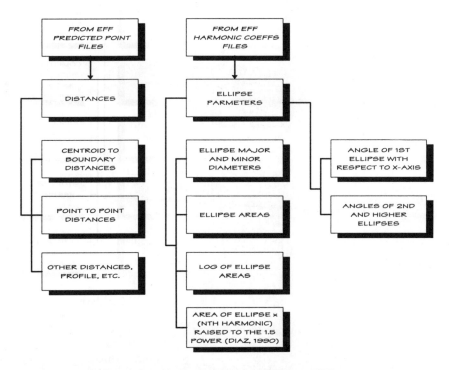

Figure A.3. Additional data extractable from EFFs.

Program Availability:

P. Lestrel, 7327 De Celis Avenue, Van Nuys, California 91406, United States, Phone: (818) 781-8499 (voice and FAX), Email: **plestrel@ucla.edu**; *or* C. Wolfe, 13376 Dronfield Avenue, Sylmar, California 91342 United States, Phone: (818) 367-6798, Email: **cawolfe@instanet.com.** Arrangements can usually be made for delivery, installation and tutorial sessions on use of the software. Contact the above individuals for details regarding supported hardware, license fees, etc. The software currently is a DOS version with *minimum* requirements of MS-DOS 3.1; an 80386 central processing unit (CPU) with a coprocessor (80387); 640 kilobytes of random access memory (RAM) with 550+ kilobytes RAM available for user programs; and a VGA color or monochrome monitor. An 80486 or Pentium CPU will prove beneficial in terms of speed if working with large samples, complex forms and/or doing 3-D analysis. A Windows 95/98 version is under development.

INDEX